Emerging Telecommunications Networks

The International Handbook of Telecommunications Economics, Volume II

Edited by

Gary Madden

Professor of Economics, Curtin University of Technology and Director, Communications Economics and Electronic Markets Research Centre, Australia

Edward Elgar
Cheltenham, UK • Northampton, MA, USA

Published by
Edward Elgar Publishing Limited
Glensanda House
Montpellier Parade
Cheltenham
Glos GL50 1UA
UK

Edward Elgar Publishing, Inc.
136 West Street
Suite 202
Northampton
Massachusetts 01060
USA

A catalogue record for this book
is available from the British Library

Library of Congress Cataloguing in Publication Data

Emerging telecommunications networks/edited by Gary Madden.
    p. cm.
   (The international handbook of telecommunications economics, vol. II)
   Includes bibliographical references and index.
  1. Telecommunication. 2. Telecommunication systems.
  I. Madden, Gary. II. Series.
  HE7631.E45   2003
  384—dc21   2002032057

ISBN 1 84064 296 3

Printed and bound in Great Britain by MPG Books Ltd, Bodmin, Cornwall

Contents

Contents of the Handbook

VOLUME I TRADITIONAL TELECOMMUNICATIONS NETWORKS

VOLUME II   EMERGING TELECOMMUNICATIONS NETWORKS

VOLUME III   WORLD TELECOMMUNICATIONS MARKETS

Part 4   Regional Developments

Figures

Tables

Contributors

James H. Alleman is a Visiting Associate Professor in the Media, Communications and Entertainment Program at Columbia Business School, and Director of Research, Columbia Institute for Tele-Information, New York and Associate Professor for the Interdisciplinary Telecommunications Program at the University of Colorado, USA. Dr Alleman founded Paragon Service International – a telecommunications call-back firm. He has been granted patents on the call-back process widely used by the industry.

Cristiano Antonelli is Chair and Professor of Economics at the University of Torino, Torino, Italy. He is Vice-President of the International Schumpeter Society, Managing Editor of *Economics of Innovation and New Technology*, Associate Editor of *Information Economics and Policy* and Co-editor of the Kluwer series *Economics of Science, Technology and Innovation*. His recent findings are contained in *The Microeconomics of Technological Systems* (Oxford University Press, 2001). He has authored papers on the economics of innovation published in *Economics of Innovation and New Technology*, *Industrial and Corporate Change*, *Information Economics and Policy*, the *International Journal of Industrial Organization*, the *Journal of Economic Behavior and Organization*, the *Journal of Evolutionary Economics* and *Manchester School*.

Vinton G. Cerf is Senior Vice-President of Internet Architecture and Technology for WorldCom, Washington, DC, USA. Widely known as a 'Father of the Internet', Cerf is the co-designer of the TCP/IP protocols and the architecture of the Internet. In December 1997, Cerf received the US National Medal of Technology. He serves as Chairman of the board of the Internet Corporation for Assigned Names and Numbers and was Founding President of the Internet Society from 1992 to 1995. Cerf is a member of the US Presidential Information Technology Advisory Committee. Cerf sits on the Board of Directors for Nuance Corporation, Avanex Corporation and CoSine Corporation. Cerf is a Fellow of the IEEE, the ACM, the American Association for the Advancement of Science, the American Academy of Arts and Sciences, the Computer History Museum and the National Academy of Engineering. Cerf holds a BSc in Mathematics from Stanford University and an MSc and a PhD in Computer Science from UCLA.

Kevin T. Duffy-Deno is a Senior Director at TNS Telecoms, a subsidiary of Taylor Nelson Sofres, in Horsham, USA. Dr. Duffy-Deno's expertise is quantitative analysis. His recent efforts have focused on the analysis of telecommunications cost models and on the development of residential and business models of telecommunications demand, with particular interest in demand models for broadband access to the Internet and for bandwidth access capacity. Prior to joining TNST, Dr Duffy-Deno was an economist with the Utah Division of Public Utilities and the Utah Energy Office, and served on the faculties of various universities. His professional papers have appeared in *Information Economics and Policy*, the *Journal of Regional Science*, the *Southern Economic Journal*, the *Journal of Urban Economics*, *Energy Economics*, and *Resource and Energy Economics*.

Scott C. Forbes is a Senior Technology Auditor at Nextel Communications, USA, where he develops risk assessments for the company's wireless telecommunications networks and mobile commerce strategies. He has been a consultant to the FCC, private telecommunications and media companies, and non-profit organizations on policy issues related to telecommunications access and online privacy. He is a Fellow of Pennsylvania State University's Institute for Information Policy. He has authored papers and presented on topics such as universal service and information security. Forbes holds a PhD in Telecommunications from Pennsylvania State University.

Robert M. Frieden serves as Professor of Telecommunications at Pennsylvania State University, USA. In addition to his teaching and research duties, Professor Frieden provides corporate training, legal counsel and consulting services in such diverse fields as personal and mobile communications, facilities interconnection, satellites and privatization. He has authored articles and books, including comprehensive primers on international telecommunications, cable and satellite television and communications law. His most recent book is entitled *Managing Internet-driven Change in International Telecommunications* (Artech, 2001). Professor Frieden is a graduate of the University of Virginia School of Law (JD 1980) and the University of Pennsylvania (BA 1977) (with distinction).

Steven Globerman is Kaiser Professor of International Business and Director of the Center for International Business at Western Washington University, Bellingham, USA. His research activities have covered a range of issues in industrial economics and international business. He has held permanent and visiting faculty positions at universities in North America and Europe and has consulted for public and private sector organizations.

Harald Gruber is Senior Economist responsible for economic analysis of the telecommunications sector at the Projects Directorate of the European

Investment Bank in Luxembourg. He also teaches at the University of Siena. His special interest is mobile telecommunications on which he recently published articles in the *European Economic Review*, *Telecommunications Policy*, *Information Economics and Policy* and the *International Journal of Industrial Organisation*. Previously he has researched the semiconductor industry, publishing articles and a book entitled *Learning and Strategic Product Innovation: Theory and Evidence for the Semiconductor Industry* (North-Holland).

Jerry A. Hausman is the John and Jennie S. MacDonald Professor of Economics at Massachusetts Institute of Technology, Cambridge, Massachusetts, USA. He is Director of the MIT Telecommunications Economics Research Program. Professor Hausman teaches a course on 'Competition in Telecommunications' to graduate students in economics and business. Professor Hausman received the John Bates Clark Award from the American Economics Association in 1985 for the most outstanding contributions to economics by an economist under 40 years of age. He also received the Frisch Medal from the Econometric Society. Professor Hausman has researched in telecommunications since 1974. He has served as a consultant for telecommunications firms and advised regulatory agencies in the US, the UK, Canada, Mexico, Sweden, Spain, Australia and Hong Kong.

Donald J. Kridel has been an Associate Professor of Economics at the University of Missouri at St Louis, USA since 1993. His primary teaching responsibilities are applied econometrics, forecasting and telecommunications economics. Earlier, Dr Kridel was the Director of Strategic Marketing at Southwestern Bell Corporation (now SBC Communications). He earned his PhD in Economics from the University of Arizona. Kridel has been active in telecommunications demand analysis and pricing research for 20 years.

Michael D. Pelcovits is Vice President and Chief Economist at WorldCom, Washington, DC, USA, where he is responsible for analysis and advocacy on domestic and international economic issues before government agencies, legislative bodies and courts. He joined MCI (which later merged with WorldCom) in 1988. Earlier, Dr Pelcovits was a Principal of the Washington, DC consulting firm of Cornell, Pelcovits and Brenner. Before entering the consulting practice Dr Pelcovits served as a Senior Economist at the FCC, and prior to that he was an Assistant Professor of Economics at the University of Maryland, College Park. Dr Pelcovits graduated *summa cum laude* from the University of Rochester in 1972 and received his PhD in Economics from MIT in 1976, where he was a National Science Foundation Fellow.

Joseph N. Pelton is Deputy Director of the Institute for Applied Space Research at George Washington University, Washington, DC, USA. Dr Pelton is also the Executive Director of the Arthur C. Clarke Institute for Telecommunications and Information that undertakes collaborative research with institutes in Europe, North America and Asia. He has served as Chairman and Dean of the International Space University of Strasbourg, Director of the Interdisciplinary Telecommunications Program at the University of Colorado at Boulder and Director of Strategic Policy for the Intelsat global satellite system. Dr Pelton has authored books including the Pulitzer Prize nominated *Global Talk*. He is elected to the Hall of Fame of the Society of Satellite Professionals and is a Full Member of the International Academy of Astronautics.

Paul N. Rappoport is an Associate Professor of Economics in the Department of Economics at Temple University, Philadelphia, USA. He is the founder and former Chairman of PNR and Associates. He is also a Senior Academic Fellow in Temple's E-Commerce Institute. His current research interests include modelling business demand for bandwidth, measuring Internet transactions, developing Internet metrics and e-commerce benchmarks.

Jorge Reina Schement is Professor and Co-Director of the Institute for Information Policy in the College of Communications and the School of Information Science and Technology at Pennsylvania State University, USA. He holds a PhD from the Institute for Communication Research at Stanford University. His recent books include *Global Networks* (1999), *Tendencies and Tensions of the Information Age* (1995) and *Between Communication and Information* (1993). His research focuses on the policy consequences of the production and consumption of information. In 1994 he served as Director of the FCC's Information Policy Project that recognized the Digital Divide. He introduced universal service as an evolving concept. He is chairman of the TPRC and is a board member of the Media Access Project, Libraries for the Future and the Benton Foundation.

Timothy J. Tardiff is a Vice President in the Cambridge office of National Economic Research Associates, Massachusetts, USA. He received a BSc in Mathematics from Caltech and a PhD in Social Science from the University of California, Irvine. At NERA since 1984, he evaluates pricing policies for competitive telecommunications markets, including incentive regulation plans and prices for access services to competitors, studies actual and potential demand for telecommunications services and develops approaches for measuring incremental costs of telecommunications services.

Lester D. Taylor is Professor of Economics and Professor of Agricultural and Natural Resource Economics at the University of Arizona, Tuscon, USA. He has a PhD in Economics from Harvard, and taught at Harvard and the University of Michigan before moving to Arizona in 1972. For the past 25 years, his published research has focused heavily on the telecommunications industry. His most recent book in telecommunications, *Telecommunications Demand in Theory and Practice*, was published in 1994.

Tommaso M. Valletti is Lecturer in Economics at Imperial College Management School, London, UK, and at the Politecnico di Torino, Italy. He is also a research affiliate of the Centre for Economic Policy Research in London. He has previously held teaching positions at the London School of Economics, where he obtained his PhD in Economics. His research interests focus on industrial economics, regulation and economics of telecommunications. He has published articles in *Information Economics and Policy*, the *Journal of Regulatory Economics*, *Regional Science and Urban Economics* and the *Review of Industrial Organization*.

Michael R. Ward is an Associate Professor of Consumer Economics at the University of Illinois at Urbana-Champaign, USA. He earned his PhD in Economics from the University of Chicago in 1993. He served as a Staff Economist at the Federal Trade Commission from 1991 to 1995, when he came to the University of Illinois. His research focuses on competition and market imperfections in high technology industries, especially the telecommunications industry. Of particular interest are the effects of imperfect information on consumers' online choices, consumers' use of information on the Internet, market institutions that have developed on the Internet to facilitate information transfer and the market consequences of consumer information gathering.

Preface

PURPOSE AND MOTIVATION

The *International Handbook of Telecommunications Economics* aims to serve as a source of reference and a technical supplement for the field of telecommunications economics. The intention is to provide both comprehensive and up-to-date surveys of recent developments and the state of various aspects of telecommunications markets. The Handbook is meant for professional use by economists, advanced undergraduate and graduate students, and also to prove useful to policy analysts, engineers and managers within the industry. With the last two decades of the twentieth century witnessing revolutionary technology and institutional change in telecommunications markets, publication of the Handbook is timely. In terms of organization it has proved convenient to categorize material into volumes on the basis of the economics pertaining to traditional 'narrowband technology' markets, markets associated with emerging 'broadband technologies' and the political economy and institutions that structure markets within which telecommunications companies operate and evolve.

The industrial organization of traditional or circuit-switched markets was primarily that of mandated monopolies offering a small array of narrowband services such as local, long-distance and international calling. This structure was supported by natural monopoly arguments. Not surprisingly, research in this era often concerned the analysis and review of production, costs and the productivity of operators. Further, the increasing privatization of incumbent operators required them to produce closer to minimum cost. Accordingly, studies examining the economies of scale and scope were prominent. Market deregulation and liberalization led to the rebalancing of rates so as to reduce cross-subsidization. In this changing environment, the analysis of competition, demand, pricing and aspects of industrial organization were important ongoing research themes. With the growth of global telecommunications, the organization of international markets and their pricing and other arrangements came under increasing scrutiny. The role of telecommunications in economic development was also considered an important discourse. During this regime welfare aspects of telecommunications markets were concerned with identifying and fulfilling universal service obligations.

In the face of developments in digital technology (mobile, satellite and terrestrial), telecommunications, cable TV, broadcasting and computers are fast converging into a single industry. The advent of these high-capacity and intelligent networks has resulted in the development of new services, such as SMS, home banking and Internet services. The proliferation of networks and their need for, more or less, seamless interconnection has led to industrial restructuring through mergers and alliances. Such reorganization is also intended to allow operators to offer a full range of services. Growth in Internet network infrastructure and subscription has provided a base for the development of e-commerce markets. Accordingly much recent research on broadband networks is forward-looking and concerned with innovation and the future of these networks, for example, the forecasting of Internet telephony adoption, the impact of e-commerce on industrial organization and the structure of future retail markets. Fixed, mobile and satellite market segments, their interaction, regulation and pricing are also matters of importance to network planners and other market participants. When faced with competing technology platforms, practically for the first time, investment decisions are more complex. Appropriate investment decisions, inherently difficult, are made more so because of the price volatility of extant technology due to ongoing innovation to enhance its productivity. In this environment new treatments of uncertainty in telecommunications are being developed. This broadband regime brings with it concerns of identifying appropriate standards and delivery for universal service.

Finally, the structure within which modern communications companies operate and evolve is complex, Accordingly, corporate strategy has to evolve beyond that relevant to a national monopoly, which may provide international service, to deal optimally with new technology and its associated service innovations, and changing composition and patterns of demand. Corporations must also account for multiple objectives associated with both national economic and social policy. Further, the subtle transition from communication to information markets places additional demands on corporate trajectories. In this light the recent corporate experience of major international telecommunications companies can provide useful lessons. Exploring the interaction of diversity in national approaches with the continuing need for international cooperation and coordination continues to be an important area of debate.

ORGANIZATION

The organization of the *International Handbook of Telecommunications Economics* has sought to reflect principal topics in the field that have either received intensive research attention or are in need of an integrative survey.

The chapters of the Handbook can be read independently, though many are complementary. Volume I begins the Handbook with three chapters on production, cost and productivity. Russel Cooper, Erwin Diewert and Terry Wales (Chapter 1) focus on the subadditivity of cost functions. To determine whether a multi-product regulated industry is a natural monopoly, it is helpful to estimate a multiple output cost function for the incumbent firm and then determine whether this estimated cost function is subadditive. This chapter considers issues in the development of functional forms capable of addressing the subadditivity question in a theoretically valid context. Yale Braunstein and Grant Coble-Neal (Chapter 2) consider issues related to cost function estimation in the telecommunications industry. Their coverage encompasses concerns associated with specification of functional forms, problems with data and approaches to addressing them, and the treatment of technical change. Estimation and testing of economic theory are also discussed. Laurits Christensen, Philip Schoech and Mark Meitzen (Chapter 3) present a survey of TFP methods and studies for the telecommunications industry. TFP is widely recognized as a comprehensive measure of productive efficiency. Further, methodological issues in measuring TFP are considered. The relationship between TFP and output growth is also examined. Finally, a discussion of TFP and its application to price cap regulation concludes the chapter.

The next four chapters of Volume I analyse the traditional areas of concern in telecommunications markets, that is, competition and market structure, demand, pricing of network services and access, and vertical integration. T. Randolph Beard and George Ford (Chapter 4) focus on competition and structure in local and toll markets. They provide a review of conceptual and empirical analyses of competition relevant to the telecommunications industry. A peculiar result they provide is that competition may increase market cross-subsidy. Competition in parts of a geographic market, or fragmented duopoly, is also formally considered. Finally, merger effects are analysed using an empirical simulation model to evaluate competition in the long-distance industry. Lester Taylor (Chapter 5) notes that demand analysts must contend with shifting industry boundaries, new services, mobile telephony growth, disappearing data sources and emergence of the Internet. He describes in detail several studies that illustrate fundamental features of telecommunications demand, with regard to both theory and technique, and transcend their particular applications. Benjamin Hermalin and Michael Katz (Chapter 6) summarize and integrate the literature on retail pricing of telecommunications network services. Their central concerns are how to price network access and the exchange of information based on use. They focus on four characteristics and their implications for socially and privately optimal pricing to end-users: consumption decisions have external effects, parties exchanging messages often have an ongoing relationship, service providers can identify customers, and incremental costs are often below average costs. Dennis Weisman (Chapter 7) considers that the dominant trends in the telecommunications industry with respect to market structure are consolidation and vertical integration. Such developments contain complex public policy and regulatory issues that both renew old debates and generate new ones. This chapter focuses on the strategic motivation for vertical integration in telecommunications markets and the corresponding regulatory and public policy issues.

The following three complementary chapters are concerned with international telecommunications markets. Douglas Galbi (Chapter 8) outlines key aspects of industry economics, firm organization and public policy relevant to understanding the globalization of telecommunications competition. Michael Einhorn (Chapter 9) reviews contemporary US policy concerning international settlements of voice and data telephone traffic. Accordingly, the FCC's previous settlements policies and the problems they posed to US carriers and callers are examined and FCC's strategy for reform is then detailed. The market under more competitive regimes is also considered. M. Ishaq Nadiri and Banani Nandi (Chapter 10) examine the contribution of telecommunications infrastructure to economic growth. Studies based on an I–O framework, and econometric cost models that describe telecommunications capital effects on industry cost structures and national productivity growth are considered. The social rate of return to telecommunications investment is discussed therein.

James Alleman and Paul Rappoport (Chapter 11) conclude Volume I with a review of the universal service concept. They argue that policy is inefficient as it is not directed to marginal subscribers, and it is costly to support because it is not targeted. Further, they consider that cross-subsidy reduces the demand for subscriber access from the group that it is intended to aid. Finally, subsidy inhibits effective competition by preventing the market from testing the efficiency of providers.

Volume II is devoted to the economics of high capacity and intelligent networks, and commences with three related but diverse chapters that cover information market innovation, Internet network history and evolution, and forecast its future. Cristiano Antonelli (Chapter 1) provides a framework to consider the accumulation of localized knowledge and competence dynamics. Such dynamics are from innovative agent activity. Agents are bounded by their competence and exposed to changing price and non-price competition. Michael Pelcovits and Vinton Cerf (Chapter 2) describe the Internet and outline its history to enable discussion of the economic forces that shape the network, and how the Internet's economics are different from those of traditional telecommunications networks. The chapter concludes with a gaze into the future of the Internet. Paul Rappoport, Donald Kridel, Lester Taylor, James Alleman and Kevin Duffy-Deno (Chapter 3) use data from an omnibus survey to empirically estimate models of consumer Internet access. They consider dial-up access only, dial-up or broadband access and broadband but cable modem or ADSL choice situations. Results confirm that availability matters. When all forms of access are available, ADSL and cable are strong substitutes to dial-up access. ADSL and cable modems are substitutes. Other factors affecting broadband adoption are the type, duration and reach of use.

The next four chapters in Volume II concern e-commerce markets and, more generally, online activity and its regulation. Steven Globerman (Chapter 4) provides a description of e-commerce, its current state and prospects. Identification and assessment of hypotheses linking e-commerce growth to changes in organizational characteristics are made, and evidence evaluated. Michael Ward (Chapter 5) distinguishes online from traditional retail markets, and asserts that lower consumer search costs lead to greater substitutability among homogeneous goods, but stronger brand preference for heterogeneous goods. Reduced delivery costs are argued to lead to rapid adoption of online retailing for travel and financial products, while higher delivery costs suggest more modest online retailing for groceries and sporting goods adoption. Robert Frieden (Chapter 6) addresses problems and opportunities presented by services provided via the Internet. After reviewing the pre-existing regulatory models that apply, the chapter considers whether a single legal and regulatory foundation exists to address the public policy issues raised by Internet-delivered communications and e-commerce. The argument concludes with suggestions on how governments should promote telecommunication and information processing infrastructure use as a conduit for services that promote education, rural development and commerce. Timothy Tardiff (Chapter 7) sees that trends in product innovation are producing complex service offerings and pricing structures. As a result, the prescription that price equals marginal cost in competitive markets is inadequate. To understand this outcome and related pricing patterns, cost structures of firms offering these services are considered and their pricing implications discussed. Examples of pricing trends for retail telecommunications services include the growth of pricing options in place of single undifferentiated prices, and the offering of 'one-stop shopping'. Finally, whether pricing responses are likely to be pro- or anti-competitive is discussed.

The following three chapters concern alternative technology platforms, their convergence and regulation, and the difficulties facing operators in making network investment decisions. Harald Gruber and Tommaso

Valletti (Chapter 8) show mobile telephony development is restricted by the available radio spectrum. Switching from analogue to digital transmission has somewhat relaxed the constraint. Further, extending cellular mobile telephony to higher frequency bands has allowed more subscribers to be accommodated. By relaxing the spectrum constraint more firms are able to supply service, and so improve service quality and price. Empirical studies find that digital technology and firm entry accelerate diffusion, but that national differences in diffusion persist. Joseph Pelton (Chapter 9) argues that satellite services growth is driven by price and technical competition, regulatory and trade reform, and advance in frequency allocation. However, satellite communications remain a complementary service and supplement to ground-based systems. Developing countries are a major growth opportunity for satellite operators by filling niches in their broadband fibre networks. He argues that geosynchronous satellite systems will dominate LEO satellite systems, and that stratospheric platforms will reduce LEO system opportunity further. Jerry Hausman (Chapter 10) regards advice on setting regulated prices is flawed when it does not recognize substantial industry fixed costs. Further, it is often claimed that economies of scale are such that regulated firms' costs are below those of entrants. Another problem arises when a regulated firm wants to decrease prices of services subject to competition. Most economists conclude that cost-based regulation leads to consumer harm and so regulatory attempts to set prices independent of demand does not make sense. Within this approach, why the failure to take account of sunk costs leads to a downward bias in regulated prices is discussed. Assumptions that network investment is fixed, not sunk, lead to error. Finally, which incumbent network elements should be subject to mandatory unbundling is considered.

Jorge Schement and Scott Forbes (Chapter 11) conclude Volume II with a discussion of universal service policy for the twenty-first century. They argue that basing policy solely on the short-term constraints faced by corporations and government misses an opportunity to build an equitable foundation for a global information age. Universal service should be sensitive to diverse population needs. The old concept of universal service simply represents an intention to wire the nation. It is proposed here that welfare is optimized when populations choose their access configuration.

Volume III contains a collection of chapters concerning the structure within which modern communications companies operate and evolve. This emerging structure is reflected in the transition from telecommunications to information social policy induced by service transformation. In this context, selected corporate experiences, international cooperation and coordination operations, and regional studies help describe the complex environment in which corporate strategies must develop and be applied. Eli

Noam (Chapter 1) describes the evolution of the telecommunications sector from incumbent monopoly status. He argues that incumbent carriers are in the midst of vertical diversification and horizontal expansion. Entrants with more efficient technology that has a greater modality impose these processes on incumbents. To maximize the gains from Schumpeterian creative destruction requires regulatory policy that ensures major companies become rivals and not partners.

The next three chapters concern the evolution of telecommunication policy to information social policy. Martin Fransman (Chapter 2) analyses the transformation from old to new telecommunications industry, and on to info-communications industry. He argues that the demise of the old telecommunications industry began with the opening of monopoly markets to competition, and the EU and WTO agreeing to liberalization. Further transformation to an info-communications industry was due to packet switching, IP and WWW technology. William Melody (Chapter 3) provides a review of telecommunications reform and information infrastructure development by focusing on traditional and network economy indicators. Traditional indicators include measures of network access and service, and competition in fixed network and mobile telephony markets. Information infrastructure development is measured by network investment, availability and use of new access technology, and Internet market development. Countries selected are leaders in global markets. Erik Bohlin (Chapter 4) considers that political and convergence dynamics will require telecommunications policy research to address a wider and more comprehensive set of questions. His chapter elaborates the nature of telecommunications policy research, addressing political inertia and its implications for policy, explaining the increased scope of telecommunications policy, and suggesting future research needs, including that of a sustainable information society.

The four chapters that follow describe the experience and visions of communications corporations. Peter Curwen (Chapter 5) argues that initial attempts to form global alliances floundered, even though 'New' Concert may have arisen from the ashes of its predecessor. Alliance formation is a slow process, partly because of the time required to obtain regulatory clearance, and because partners need time to ensure network architecture and standards are compatible. Time is a scarce resource in a rapidly changing industry. Furthermore, alliances consistently underestimate the difficulty of keeping relatively loose arrangements in place. Niall Levine, Douglas Pitt and David Lal (Chapter 6) analyse changes BT faced in the 1990s. They argue that the causes and effects of change are relevant to the disciplines of strategic planning, organizational behaviour, telecommunications management, economics and regulatory policy. They contend that BT faces permanent revolution in redefining its strategy and organizational practices to meet emerging market opportunities and environmental challenges posed by increasing competition and regulation. Martin Taschdjian (Chapter 7) considers that US WEST's evolution into international markets and broadband technology is typical of the telecommunications revolution. Its history of strategic initiatives, and the learning and adjustments made, offer insight into the issues of convergence, diversification, technology, competition, shareholder value and public policy. These lessons suggest AT&T's strategy of evolving from a long-distance carrier into a local broadband services company faced significant challenges. James Alleman and Lawrence Cole (Chapter 8) conclude that GTE purchased the long-distance telephone company SPC because it was already involved in telecommunications and other high-technology business, had an acceptable investment return, and an effective management in handling regulatory process. However, they claim the move was imprudent. Important mistakes include an inadequate understanding of interconnection charges and valuation analysis, and of a government imposed Consent Decree.

The seven chapters that appear next provide a complementary overview and analysis of national and international agencies that affect regulatory policy and international telecommunications aid activity. Richard Cawley (Chapter 9) provides insight into the substance and process of telecommunications regulation and policy in the EU in the latter part of the twentieth century. He considers an important message in the EU context is that regulatory policy can often not be designed from a first-best standpoint. It is constrained by the legal and institutional structures that are enshrined in successive EU treaties, and these in turn reflect evolving political, social and economic prerogatives and compromise between member states. Timothy Brennan (Chapter 10) argues that across the US, cable TV systems have begun to expand offerings beyond multi-channel point-to-multipoint video to include high-speed broadband Internet service. Recently, the US Federal Trade Commission, with the FCC, imposed open access requirements on Time Warner's cable systems as a condition for approving its acquisition by America On-Line. The merits of imposing such broadband access requirements are debatable. To ensure locality access rules are efficiency based, open access requirements should be imposed independently of merger occurrence. Tim Kelly (Chapter 11) asserts that ITRs, among the longest surviving international treaties, are regarded, especially in developing countries, as an expression of national sovereignty. The current ITRs, adopted in an era when old certainties were tumbling, remain essentially unchanged. For example, at the time of drafting there were only half a million Internet users globally. The current situation is one of atrophy in which the ITRs are slowly falling into neglect. Dimitri Ypsilanti (Chapter

12) considers that regulators are trying to interpret principles that are largely empty of detail. An associated difficulty is gaining consensus on the degree to which the details of policy should be the same, as opposed to the general frameworks. A means to ensure the continued convergence of regulation is through sharing experiences, and more open evaluation of regulator performance including benchmarking. Such an approach will ensure best practice regulation is more widely adopted and provide an impetus for convergence in policy and the implementation of regulation. Bruno Lanvin (Chapter 13) addresses the question as to whether it is too late to bridge the digital divide by stressing the main aspects of the present approach. That is, how do the attitudes of stakeholder groups differ from what they were two or three years ago? Why has the issue recently moved from an important problem to a matter of emergency? Finally, how are the nations of the G-8 gearing up to react to this challenge? Charles Kenny (Chapter 14) provides an early history of WBG involvement in the sector and describes change in the 1990s. In particular, the IFC undertook the role of investing in private telecommunications and the WBG moved toward supporting reform in policy and the development of regulatory institutions. After discussing developments in the late 1990s, including early reaction to evolution of the Internet, the increasing importance of universal access and the need for broadened strategy focus, the chapter examines ongoing efforts to reform the WBG strategy for information infrastructure. Claude Barfield and Steven Anderson (Chapter 15) indicate that 50 per cent of WTO members made no basic telecommunications commitment to the GATS. Further, they argue that ongoing negotiation should insist countries meet their 1997 commitments. Negotiations should also harvest more commitments from informal unilateral liberalization. An obvious omission involves basic telecommunications and technology convergence. Future negotiation must gain commitments concerning technology and anticompetitive practice by non-telecommunications entrants. Finally, the WTO basic telecommunications agreement needs to confront international pricing.

The last eight chapters in Volume III are concerned with regional issues. It is hoped that while the issues may be peculiar their lessons will be widely applicable. Andrea Kavanaugh (Chapter 16) considers that any assertion that the Internet favours a new pan-Arabism is ill founded and a more likely outcome is increased pan-elitism. Wealthy, well-educated and young Arab new media users are improving their lifestyles through consumerism, entertainment and realistic news coverage. Nor does the Internet suggest there will be new pan-Arabism, as elite Arabs are drifting farther from their less advantaged brethren. The gap between the haves and have-nots in the Arab Middle East is widening at a great pace. Mark Shadur, Kellie Caught and

René Kienzle (Chapter 17) argue that Australia and New Zealand have major differences in their telecommunications sectors as a result of deregulation. These differences have resulted in distinct market structures, competitor strategy and user outcomes. The chapter outlines change programmes for organizational design, work organization, employee involvement and SHRM. The question as to whether organizational changes have improved operational outcomes remains. Marcelo Resende (Chapter 18) traces the recent evolution of the Brazilian telecommunications sector, in particular the public provision of telecommunications and the system that replaced it. His principal focus is the emerging regulatory framework and challenges facing the sector. The period prior to privatization is examined through aggregate and firm-specific productivity indicators, and a description of the main events defining the institutional design is given. An examination of the post-privatization environment follows. Meheroo Jussawalla (Chapter 19) posits that as Chinese leadership encouraged the deployment of the latest IT and pursued development of global information infrastructure through the mid-1990s, the existing telephone network became obsolete and led to a planned broadband cable network. China now has to devise a future to better diffuse the information superhighway. Further, China needs to combine the goals of information infrastructure advancement with investment in primary education or face social unrest. Nicolas Curien and Dominique Bureau (Chapter 20) describe the change in regulatory oversight of the French telecommunications and electricity sectors. Among network industries, telecommunications and electricity are not uniquely affected by liberalization and market-oriented EU directives. Given the criteria of opening markets to competition and maintaining the public operators' dominant position in certain market segments, telecommunications and electricity are interesting sectors to study. Moreover sector comparisons are relevant as the public sector companies involved, France Télécom and EdF, are efficient and able to withstand international competition. Günter Knieps (Chapter 21) considers the German entry process and its relationship to EU network liberalization policy. New German telecommunications law tends toward over-regulation and a large regulatory base. Further, the potential for phasing out sector-specific regulation is sketched. A critical appraisal of the Review of the German Monopoly Commission is also provided. The analysis shows that after entry deregulation, regulation of market power is only justified in local telecommunications networks when they constitute monopolistic bottlenecks. Elsewhere actual and potential competition is sufficient to discipline the market. T.H. Chowdary (Chapter 22) observes that until recently the Indian telecommunications sector was characterized by millions of applicants waiting for network connection, unsatisfactory service delivery, inflated telephone tariffs – too high to generate funds for network improvement – inability to develop technology locally and poor communications for export-orientated business. Automation and computerization of telephone directory assistance, billing, complaint and fault registration, and control to improve quality of service were resisted. He describes how the move from centralized planning, liberalization, accession to the WTO and signing of the ITA have profoundly affected the provision of telecommunications and information services. Willem Hulsink and Andrew Davies (Chapter 23) argue that national governance regimes shape the internationalization approach of European PTOs. The strategic choice of PTOs, rivals and regulatory agencies determine the scope of market restructuring. The economics of innovation helps to explain innovation strategy. Occasionally industry is altered by strategic behaviour, disruptive technology and exogenous shocks, and so challenges the political and economic system. These situations create dissonance and technology may be replaced, and winning corporate strategy and industry dynamics transformed.

Acknowledgements

My primary obligation is to the authors of the chapters that comprise the *International Handbook of Telecommunications Economics*. I appreciate the authority, talent and imagination with which they prepared their drafts, the willingness to accept comment and efforts to meet deadlines.

Gary Madden
Curtin University of Technology

1. Innovation in advanced telecommunications networks

Cristiano Antonelli

INTRODUCTION

Advanced telecommunications provide an interesting area of analysis for the economics of innovation.[1] The rate of technological change in advanced telecommunications since the late 1950s has been rapid. This technical change has led to general-purpose technological systems with a wide scope for application. Further, technological change is characterized by the introduction and market selection of an array of technological innovations, each of which exhibits high levels of complementarity and interrelatedness with different technologies. Such process and organizational innovation also changes the production mix of firms and the composition of their markets. This array of technological innovations is characterized by the complementarity of innovative efforts of a variety of players, each with its own localized field of activity. The emergence of such a cluster of complementary innovations appears to be the outcome of the sequential introduction of interdependent innovations along definable paths shaped by a number of factors. They include complementarities (technological externalities and economies of scope), advantages of learning in specific areas of application; competence accumulated by the dynamics of learning to do and use; and the irreversibility and indivisibility associated with significant portions of the material and immaterial capital structure of existing firms.

The next section presents an interpretative framework and considers the role of super-fixed productive factors and the dynamics of accumulation of localized knowledge and competence as the key focusing devices for understanding the market construction of technology change (Appendix A presents the formal model). Following that, the dynamics of technological change in telecommunications markets are outlined and contrasting interactions in the generation of innovations are described. The major technological changes in telecommunications markets are itemized in this section, and their explanation in terms of the market construction of technological innovations illustrates the applicability of the model used (Appendix B presents a tentative classification of the main technological innovations recently introduced in telecommunications markets). In the conclusion, the dynamics of technological change are presented as the outcome of the innovative efforts of agents bounded by their competence and their history, and exposed to the changing conditions of competitive markets characterized by both price and non-price competition.

MARKET CONSTRUCTION OF NEW TECHNOLOGY

The economics of innovation is facing the emergence of a new approach to understanding the micro-dynamics of technological change, which parallels the diffusion in both the history and the sociology of science and technology of the social constructionist school (Bijker, 1987; Latour, 1987; David, 1992, 1993, 1994; Misa, 1995; Smith and Marx, 1995). Social constructionism, as articulated by Latour (1987), contrasts with the deterministic model based on technological trajectories, and stresses the role of the collective undertakings of a myriad of innovators. The generation and introduction of technological innovations are now viewed as the result of complex alliances and compromises among heterogeneous groups of agents. Agents are diverse because of the variety of competencies and localized knowledge they build on. Alliances are based on the valorization of weak knowledge indivisibilities and local complementarities among technological knowledge. The convergence of the efforts of a variety of innovators, each of which has a specific and yet complementary technological base, can lead to the successful generation of new technology. In this approach the distinction between innovation and diffusion is blurred. The spread of adoption is viewed as the result of a complementary effort that makes useful and specifically reliable a new technology, increasing its scope for application. Adopters are no longer viewed as passive and reluctant prospective users, but rather as screeners who assess the scope for complementarity of new technology with their own specific needs and contexts of action. Profitability of adoption is the result of a process rather than a given. A technology diffuses when it is applied to a variety of diverse conditions of use. The intrinsic heterogeneity of agents applies not only to their own technological base but also to the product and factor markets in which they operate. The vintage structure of their fixed costs and tangible and intangible capital can be portrayed as major factors of differentiation and identification of the specific context of action, with respect to both technological change and market strategy. New ideas can be implemented and incrementally enriched, so as to become eventually profitable innovations, when appropriate coalitions of heterogeneous firms are formed (Antonelli, 1995, 1999).

The analysis of industrial dynamics provides an arena in which the market construction of new technology can be observed. A variety of firms can be analysed in the course of their competitive strategies where technological change plays a major role. Each firm is characterized by a specific competence on which localized knowledge is implemented. Firms are also characterized by specific endowments of fixed capital, and the contexts of their conduct in both factor and product markets. Changing market conditions induce firms to innovate in both market strategy and technology. Technology choices concerning the introduction of product and process innovations, the adoption of new technologies provided by suppliers and the imitation of competitors are mingled with market strategies such as specialization, outsourcing, diversification, entry and exit, merger and acquisitions and internal growth. In a continual trial in the market place firms experiment by changing their mix of technology and market conduct. At the aggregate level the result is the market selection of new better technologies, often characterized by strong systemic complementarity.

At the microeconomic level the rate and direction of technological change can be viewed as the endogenous outcome of the innovative sequential reaction of firms induced by the interplay between irreversibility of their capital stock and economic entropy. Irreversible and clay capital stock can be thought to be constituted by both fixed physical capital and competence and technological knowledge in well-defined and circumscribed technical fields. When irreversibility matters all changes in current business require some adjustment costs that are to be accounted for. In this approach firms are portrayed as agents whose behaviour is constrained by the irreversible and clay character of a substantial portion of their material and immaterial capital. Moreover the conduct of firms is affected by bounded rationality, which implies strong limits to their capability to search and elaborate information about markets, techniques and technology. As a consequence competence constitutes the basic irreversible production factor. In turn competence is embodied both in the organization of the firm and in the stock of fixed capital. Hence the production function of firms can be characterized by both irreversible inputs, that is super-fixed production factors and flexible production factors (Antonelli, 1995). Clearly, firms cope with changes in market demand, and in the relative prices of productive factors, only after dedicated resources have been applied to search for new and convenient routines. Consequently, firms make sequential choices reacting to a series of changes in their business environment. Notions relevant to this context are switching costs and localized knowledge.

Switching costs are defined here as the costs of changing the actual state of the production process within a given technology, because of indivisibilities, interrelatedness and non-compatibility among inputs in different dimensional mixes. The consideration of the technical features of long-lasting capital stock embodied in machinery, buildings and skills of personnel provides evidence and theoretical support to this analysis of switching costs. All changes in the levels of inputs require the necessary compatibility of additional levels of inputs to the pre-existing ones if the firm needs to expand output or to change the mix of inputs. Switching costs are important when, because of the specific limitations imposed by the characteristics of the production process, firms find it difficult to change the levels of super-fixed inputs they currently use and prefer to keep using the same amounts of inputs. These circumstances seem to be especially strong in the telecommunications industry, where the production process is characterized by the strong irreversibility and rigidity of a variety of specific inputs, often marked by indivisibility. In these circumstances short-term production and cost theory apply to a longer time span than that usually assumed. The rising portion of the short-term average cost curve defines the amount of switching costs firms are exposed to when faced with unexpected changes in their production process. Changes in the production mix and size of output expose firms to relevant price and output Farrell (1957) inefficiency, with a sharp decline in the efficient use of productive factors.

Firms are required to cope with the dynamics of demand and factor prices by introducing technological innovations and making adjustments according to market fluctuations, yet retaining as much as possible the previous levels of super-fixed inputs. Hence, technological change is introduced according to the relative costs of doing so. The introduction of new technology is not free and requires the investment of dedicated resources to conduct R&D activity, to acquire external knowledge and take advantage of new technological opportunities, and to elaborate on tacit learning processes to generate localized knowledge. Localized knowledge consists of the accumulation of the benefits of experience and learning by doing, learning by using, learning by interacting with consumers, and learning by purchasing. Firms are able to upgrade their existing technology only when they can blend the generic knowledge made available to them by new scientific discoveries and general movements of the scientific frontier with their own technological know-how. Hence localized knowledge makes it possible to capitalize on generic knowledge with the know-how acquired using the techniques currently in use. This dynamic leads firms to retain techniques close to those in current use, and so continue to improve on this technology.

When switching costs are important and firms are reluctant to change the level of inputs employed in production, especially super-fixed inputs currently used, they may prefer to keep using the same amounts of super-fixed inputs by means of the introduction of technological innovations and produce a larger output, and/or keep the pre-existing mix of inputs. Hence firms can adjust to a change in their business conditions without altering the level of super-fixed inputs. This is more plausible when the introduction of technological change is made possible by new technological opportunities together with the accumulation of competence and localized knowledge within the firm. With high switching costs and major technological opportunities – as is the case with advanced telecommunications – firms will mainly cope with economic entropy (demand fluctuations and changes in the relative price of production factors) by means of the introduction of innovations (and experience localized TFP growth) and a reduction of average costs to below the minimum level obtained with perfect flexibility of all productive factors. At this point the firm will increase the size of its output or adjust the production mix, along a path defined by the given and irreversible levels of the super-fixed factor and competence base.

Economic entropy, in these conditions, pushes both the rate and the direction of technological change. Firms, facing a continual change in business conditions, are induced to generate a stream of technological change along a well-defined path in terms of the complementarity of new technology with existing super-fixed productive factors. The new localized technology will be more intensive with super-fixed productive factors and the localized technological competence. Firms are induced to innovate by the irreversibility and economic entropy to search locally for new technology. The direction of technological change will be influenced by the search for new technologies that are complementary with existing technology. The rate of technological change is influenced by the relative efficiency of the search for new technology, for given levels of sunk costs and related switching costs. The larger the technological opportunities, the larger will be the amount of new technology firms and the economic system can generate. This dynamic leads firms to remain within a region of techniques close to those in current use, and to continue to improve on this technology. This outcome is more plausible when the introduction of technology occurs through the accumulation of competence and because of localized knowledge contained within the firm.

The dynamics of economies of scope are of central importance in this context. Economies of scope cannot be considered to be a windfall or exogenous, but rather the actual result of the introduction of new technology that is complementary with existing technology. The direction of technological change pursued by each firm leads to the search for and eventual introduction of new technology that makes use of existing portions of the capital stock and relies on the competence acquired in using the existing technology. Technological change in the telecommunications industry is usually the result of technological strategies of different groups of firms, each pursuing their own market strategy building on their localized and specific technological capability. New market entrants have a tendency to exploit new technological opportunities in order to enter the markets for telecommunications services, thus introducing centrifugal innovation, which gives them market access based on dynamic economies of scope. Such economies arise through application to existing capital stock and localized technological competencies. Incumbents entrenched in their own geographic and product markets have a clear interest in introducing technological innovation in a way that makes best use of their existing network and related localized knowledge. The changing of coalitions between different groups of players is shaped by the rate and direction of technological change.

DYNAMICS OF TECHNOLOGICAL CHANGE IN TELECOMMUNICATIONS

Advanced telecommunications provides a wide scope of applicability and elaboration across a broad range of uses (Bresnahan and Trajtenberg, 1995). Technological change can be used in a wide variety of products and processes and also has strong complementarity with existing and potential new technology.[2] For example, the convergence of computing, telecommunications and broadcasting can be considered the source of much radical technological change in the telecommunications industry. Such technological convergence is, however, the result of a complex interaction of a variety of technological and market strategies elaborated by a variety of players in the market place. From the demand side, the advent of computer communication is seen as the major factor driving the shift toward increasing the demand for telecommunications services. This rapid increase in demand for telecommunications services has pushed firms to explore new growth strategies and to search actively for new technology within a limited technological range of options as close as possible to that embodied in irreversible super-fixed productive factors. Differentiated technological strategies have progressively emerged clustered around the fixed network endowment of incumbents, the internal communications structures of large and innovative users, and radio communication opportunities.

The development of the general technology is the result of the continual selection in the market place of the rival technology introduced by an array of heterogeneous competitors. The dynamics of technological change in advanced telecommunications is best understood as the result of an ongoing technological rivalry among technological innovations. Each innovation is implemented by different groups of players with different endowments and market strategies. During the last 30 years technological change has been rapid and radical in telecommunications, especially for telecommunications transmission, switching, distribution, signalling and network management. Further, such innovation has changed the shape of the network and the complementarity among components of the network. Technological change has affected not only the market position of operators in the network but also the technical and organizational features of the architecture of the network. Each of these innovations deserves to be analysed as the outcome of the localized effort of well defined groups of actors, changing their technology and consequently their role within the network, their markets and their profitability.

The important technological innovations in the telecommunication industry can be classified into two clusters. Centripetal technologies enhance the relevance of economies of scale, scope and density, stemming from the existing super-fixed capital infrastructure constituted by the existing network. These technologies are mainly introduced by network operators in order to enhance the dynamic efficiency of their actual production mix, and more specifically of the super-fixed productive factors such as the existing network, articulated in cables, switching and transmission equipment. Other centrifugal technologies reduce the relevance of technical economies of scale and density: specializing technology reduces the role of inter-functional telecommunications economies of scope, and segmental technologies reduce the role of network externalities. These technologies are mainly introduced by large users and new mobile competitors that rely on wireless communication as a way to reduce the strong competitive advantage of incumbent network operators.

The mapping of major technological changes introduced in the telecommunications industry during the last 40 years into an economic model makes possible an attempt to understand the directions of technological change in this area as the result of the contrasting innovative efforts of a variety of players characterized by their localized search for new technologies.[3] The cluster of technological innovations introduced in the 1950s and 1960s reflected the market conditions of the telephone industry resulting from conditions of earlier investment from the 1920s through to the late 1960s, that is, natural monopoly. This innovative behaviour reflects the 'centripetal' efforts of the natural monopolist to reproduce and extend the conditions of the monopoly by the introduction of innovations that reinforce the dynamics of increasing returns. This technology path was clearly shaped by an effort to increase the intensity of use of the existing super-fixed capital infrastructure centred on the existing network. Research strategy was directed towards the switching and transmission functions. In the early 1970s, along with the emergence of new information technology, the direction of technological change in telecommunication services entered a new phase fuelled by the emerging relevance of dedicated services for large users and market opportunities in mobile telephony. This phase made the institutional set-up and established organization of the industry progressively outdated. This new cluster of technological change from the late 1960s until the mid-1980s had all the characteristics of a localized process of innovation led by large advanced users and new mobile competitors. The focus of technological change was clearly directed towards new transmission systems based on radio links aligned in the effort to minimize the requirements of Hertzian space, and to reduce the limitations of the bandwidth. Signalling technology was also an important area of activity as it was better placed to direct the increasing flows of mobile traffic. Space technology, in this context, made its earliest entry to this strategic area of civil application.

Data communication has also played a key role in this context. Initially, data communication was concerned with limited groups of very large users with few network externalities. This traffic was highly sensitive to the volume of long-distance tariffs. Attention was drawn to these markets because of the opportunities for cream skimming built into markets characterized by network externalities. Segmentation of demand into niche markets, and the identification of customer classes with few demand externalities *per se*, without any effects on the cost side, offered important market opportunities. New communication media such as dedicated data communication systems made it possible to better identify the needs of groups of customers. A large customer base, high tariff levels, and the opportunity to reap quasi-rents from the innovative usage of telecommunication services, may be considered among the strongest factors influencing the enhanced rate of introduction of technological innovations since the late 1960s. It also induced the shift in the direction of technological change following the introduction of centrifugal innovations that eventually led, at least in the US, to the segmentation and specialization of the centralized network into a web of special-purpose networks. The fast diffusion of intranets within large corporations is a major centrifugal force driving the creation of a variety of dedicated networks not needing interconnection. The large endowments of electronic hardware of large corporations and their acquired competence in managing application software played a major role. Data communication was initially introduced by large multinational companies and eventually diffused to large corporations to coordinate their global and multiple site operations by means of intranets characterized by a limited number of connected users. Idiosyncratic standards, tailored for internal communication and computing needs, also played a role. Traditional telecommunications could not easily match the variety of dedicated standards users. The architecture and management of such intranets were provided internally. Network externalities were irrelevant and telecommunications carriers just provided transmission capabilities. High telecommunications tariffs, especially on international routes, forced multinational companies to minimize the costs of their intranets by using dedicated lines. Irreversibility here acted as a basic sorting device to generate new economies of scope. The existing stock of competence in software, hardware and in managing complex data bases was readily extended to the new and innovative communication functions.

Divestiture in the US and massive privatization and liberalization in Europe provided the institutional context which allowed the entry of specialized competitors and the diversification of large corporations in telecommunications services. The centrifugal effects of technological change and industrial dynamics were also relevant in the 1980s with respect to the strong vertical ties between carriers and hardware manufacturers. In the 1980s vertical disintegration between telecommunications service providers and telecommunications hardware manufacturers took place. These vertical links were put under strain because of the increasing variety of the customer base, especially specialized newcomers and large users. At the same time, on the supply side, new technology was often supplied by market entrants. Hardware market entry led to a wave of mergers and acquisitions and resulted in the severing of traditional user–producers relations. Gradually, through the 1980s, intranets developed to become extranets. Large corporations realized the scope for application of their own internal data communications networks to manage their distribution systems and relations suppliers. Huge business-to-business data communication flows were growing in order to make possible the integrated logistics required by just-in-time management techniques. Suppliers were required to provide timely intermediate inputs. Deliveries were planned and inspected via data communication. Subcontracting was increasingly implemented with data communication systems based on more sophisticated extranets. Co-engineering and shared R&D activities were also implemented online, with a growing number of parties involved. The online management of financial services experienced exponential growth. This evolution had major consequences for the architecture of telecommunication networks. Interoperability of standards became a central issue. Internal switches had to face growing traffic. Telecommunications carriers could take advantage of their competence in managing the increasing complexity of the webs of extranets. Specialized networks declined and traditional telecommunications carriers learnt to supply bundles of dedicated services, tailored for the needs of intranets and extranets of large corporations, especially when multinational, with a global range of operations.

The social need for an advanced universal network became evident again after years of neglect. The dynamics of network externalities was again in place, with data communication the driving force. The speed and quality of digital communication became a central issue. The capillarity of extranets and their varying configurations with flows of entries and exits reduced drastically the viability of dedicated and personalized networks. Telecommunications carriers could now attempt to merge the growing traffic engendered by data communication and traditional voice telephony. Since the mid-1980s, technological change has been progressively shaped by such centripetal and integrating technological innovations as ATM, BISDN and ISDN. The application of digital technology made it easier to find centralized solutions not only to the management of administrative problems (such as billing), but also to technical problems related to traffic management (switching, routing and so on), and the supply of advanced services.

The interplay between centrifugal and centripetal forces in shaping the direction of technological change in the telecommunications industry has been observed not only in the transition between voice telephony and digital communication, but also in the duel between mobile and fixed telephony. In the late 1980s mobile telephony became the technological axis on which much technology change has been aligned in an effort to make entry easier and to minimize the need for interconnection with the existing fixed networks. Mobile telephony has been the main tool for telecommunications market entrants to overcome the substantial entry barriers existing in traditional voice telephony markets. Mobile telephony enabled the entry of competitors in newly liberalized telecommunications markets. Cross-entry of incumbents in markets was based on mobile telephony in Europe. Mobile technology has been subject to incremental innovation, paving the way for a completely new technological system based on the systematic use of Hertzian space as the main communication infrastructure. It is instructive to recall, however, that the basic technology has been available for more than 60 years; for example, mobile telephony had been used by police, armed forces and by transportation companies for many years.

The introduction in the early 1990s of the Internet and the World Wide Web marked a major shift in the direction of technological change, in that it paved the way for mass usage of the fixed network by households and related increased interest by manufacturing and service firms to interact online with a large base of customers. The advent of electronic commerce induced a major shift in the use of data communication from business-to-business communications flows to business-to-consumers flows, changing drastically the role and scope of economic interest of the existing distribution network. This trend was further fuelled by the introduction of new signalling and compression technology (such as ADSL), which made it possible to obtain greater economies of density and scope from the existing coaxial cable network. This was made possible by the ability to carry data and images at high speed. Similarly, the advent of digital TV also provided new ways to make use of the existing infrastructure for the distribution of entertainment and specialist TV programmes.

Competition between content providers and communication carriers emerged because of new opportunities that arose in attempting to take advantage of existing coaxial cables specialized in TV traffic. Online service providers tried to establish their own networks. ISP developed with the awareness of the huge potential of electronic commerce and the high costs for telecommunications services as an intermediate input. Content providers entered the industry and were induced to create their own long-distance networks to save on high interconnection tariffs. Fibre-optic technology provided an opportunity for entrants, while ADSL supplied an important opportunity for diversification to cable TV operators. A wave of mergers and acquisitions took place in the telecommunications industry, fed by the cross-entry of specialized carriers in fixed voice, mobile and data markets. The introduction of major centripetal technologies paralleled this process. Internet Protocol and UMTS in mobile telephony characterized this new phase of technological convergence, which brought together voice, mobile and data communication into a single network. Once again institutional change, especially in the US, with the Telecommunications Act in 1996, paved the way for multimedia competition leading to the creation of multipurpose networks. Emerging multipurpose telecommunications networks are now the platform for the distribution of a variety of products, including financial and entertainment services. Economies of scope in multiservice networks are the main tool for incumbents to build a competitive advantage for content-providers eager to integrate vertically into bulk transmission, while taking advantage of interconnection rights made available by regulators. This dynamic of technological convergence and divergence is leading the industrial organization of telecommunications markets toward new structures where voice communication is increasingly migrating towards mobile communications systems, and fixed telephony is becoming the central infrastructure to deliver data communication, Internet services and digital TV.

CONCLUSIONS

The direction and the rate of technological change are largely endogenous, due to the behaviour of firms and to the selection process built into the competitive arena. Technological change is introduced into telecommunications markets by firms facing an altered economic environment. Such developments incur additional costs (in terms of the search for alternative techniques), such as the re-education of employees and additional capital investment. Switching costs are important when the production process of firms is shaped by irreversibility, indivisibility and the non-malleability of capital stock, which in turn can be both material and immaterial. Such conditions restrict and make expensive movements along the existing technical isoquants, and firms are induced to explore the scope for introduction of new technology. Localized search for better technology is induced by the need to cope with changes in the economic environment, and is bounded by the competency and knowledge contained in specific routines, products and processes. Firms are able to upgrade their existing technology only when they can blend the generic knowledge made available from scientific discovery with their own technological know-how. Hence localized knowledge allows the firm to capitalize on localized generic knowledge acquired from the use of existing techniques currently in use. The outcome is the introduction of localized technological change, which results in a limited technology space gravitating around the constraints imposed by the characteristics of localized competence and technological knowledge.

When innovation is induced by the irreversibility of productive factors, which are characterized as super-fixed capital stock, new technology that is both compatible and complementary with the technical features of the existing capital is more likely to be adopted. In these circumstances a dynamic process is likely to take place where, for any demand, firms introduce localized, and hence complementary, innovations. That is, firms select technology to make better use of the existing sunk production factors, or take advantage of economies of density and create economies of scope. The outcome of the repeated interactions between demand pull and technological opportunities is best understood only at the microeconomic level. Microeconomic analysis makes it possible to go beyond the debate between demand pull and technological push as the forces that drive technological change. In the approach employed here, markets act as economic and technological laboratories where different technology embedded in different firms is searched through. The pairing of a technology and a firm can be thought of as the outcome of an experimental trial in the market place. The dynamics of localized technological knowledge, the search for complementary innovations, the creation of dynamic economies of scope and the market selection of the new technologies which take better advantage of sunk capabilities and capacities lead to the gradual emergence of new technological systems characterized by important complementarities, compatibilities and interoperability among alternative technologies.

In developing a microeconomic approach to understanding the market construction of technological change, innovations are interpreted as the result of the business strategies of different groups of firms each characterized by their competencies and capital endowments. The co-evolution of industrial dynamics and localized technological change here plays a strong role and values the continual interplay between market and technological strategy. Incumbents try to resist the entry through the introduction of centripetal technology that builds on their super-fixed tangible and intangible infrastructure. Entrants, relying on their different mix of economies of scope and density, try to build competitive advantage by introducing centrifugal and specialized technological changes to enhance the economic value of different tangible and intangible fixed capital, mainly based in electronics and service industries. The market construction of technological change is the end result of such an ongoing process.

This interpretative framework has proved to be a useful device in understanding the dynamics of technological change in advanced telecommunications markets. Such change is seen as an emerging battle of technological innovations, whose adoption is based on the direction shaped by the search for complementarities and compatibility between new technologies and the existing sunk structure of capital stock of incumbent fixed network operators and large new mobile telephone operators, respectively. The evidence of advanced telecommunications suggests that the dynamics of technological change is fuelled by the interaction between the localized technological base of a variety of heterogeneous firms and the industrial dynamics of innovative entry of new competitors, and the exit, specialization, diversification and vertical integration of incumbents so as to define an architectural and structural change whereby the reconfiguration of existing product and process technologies and the failure of either established firms or new competitors are strongly interdependent.

APPENDIX A: FORMAL ANALYSIS

To clarify the argument presented above, consider the case of a firm producing output Y_0 with the inputs I_0. A shift in the demand curve requires the firm to produce the quantity Y_1 and inputs to I_1. However, the firm is able to expand output to the level Y_1 without increasing inputs by shifting the isoquant inward and increasing the level of efficiency from a to a_0. Specifically the production function is assumed to be:

$$Y = af(F, I) \tag{1.1}$$

where a reflects the efficiency level, F is super-fixed productive factors and I is a bundle of flexible inputs. Relative prices of productive factors are assumed fixed, so that there is no need to specify the composition of the bundle of inputs and their relative efficiency. Also market prices are assumed constant.

An increase in demand permits an increase in revenue and the entry of new customers so that output should grow from Y_0 to Y_1. Output cannot be increased along the traditional long-term growth path in moving from the old to the new equilibrium. Such movement implies the perfect divisibility of the super-fixed productive factors. In such conditions, in order to accommodate the increased output, the firm follows the traditional alternative path by increasing the flexible inputs so as to be able to increase output up to the new desired level, yet retaining the given fixed inputs. The new solution, however, is clearly price inefficient. It lies to the right of the intersection between the isocost and the amount of super-fixed productive factors available.

Such a growth strategy can be expressed as:

$$Y_1 - Y_0 = af(I_1 - I_0). \tag{1.2}$$

In order to increase the output through increased levels of flexible inputs from I_0 to I_1, holding super-fixed factors constant, the firm has to cope with a clear loss of efficiency. The result from using the wrong mix of productive factors is too many flexible inputs and too few super-fixed inputs. The firm exhibits Farrell (1957) price inefficiency. The positive slope of the average cost curve reflects the excess increase in the level of flexible inputs. On this basis:

$$I^* = I_1 - I_0 = g(S) \tag{1.3}$$

where S is the resources necessary to acquire additional flexible inputs, to integrate them in the production process, and is measured by the more than proportionate increase in total costs due to the wrong mix of production factors. We assume a decreasing marginal productivity: hence, $dI^*/dS > 0$ and $dI^{*2}/dS^2 < 0$.

As an alternative to changing the level of inputs, the firm can change the general level of efficiency of the production process, and obtain more output with the current level of inputs:

$$Y_1 - Y_0 = h(TC). \tag{1.4}$$

In order to generate greater efficiency, however, better technology (TC) has to be introduced into the production process. Better technology results from innovation activity, that is, the overall amount of resources devoted to R&D, and to capitalizing on learning opportunities (RDL). All increases in output levels obtained with the same levels of inputs are the result of some innovation activity:

$$TC = b(RDL) \tag{1.5}$$

where TC measures the introduction of better technology, which would make possible an increase in output without changing the level of inputs. Here it is convenient to assume that diminishing marginal returns characterize the innovation function, so that $dTC/dRDL > 0$ and $dTC^2/dRDL^2 < 0$. In sum, adjusting the levels of inputs implies some switching costs that add to the costs of the inputs, and retaining the present level of inputs but changing the general level of efficiency by means of better technology incurs a cost of innovating.

Selection of the appropriate behaviour depends on the relative efficiency of resources invested in input-augmenting-cum-switching and innovation activity, respectively. The advantages of augmenting the level of outputs to cope with new demand via an increase of TFP rather than an increase of input, moreover, are also affected by the specific conditions of market structures and appropriability regimes. With high barriers to entry and strong appropriability, such as those prevailing in the telecommunications industry, the firm can fix prices with a constant mark-up on costs. Appropriability conditions make it possible for innovators to retain an important fraction of the increased TFP associated with the introduction of better technology. The advantage of adjusting flexible inputs to meet the increased demand is measured by the revenue, including mark-up, generated by the additional quantity:

$$Z = M(Y_1 - Y_0)/Y_0 \tag{1.6}$$

where Z is the unit net revenue from adjusting output to demand by means of an increase in inputs and M is the mark-up.

Conversely a firm may be able to expand its output yet retain its inputs fixed by introducing innovations that increase TFP. This has a benefit that can be measured by the product of the new output by the same mark-up as the firm that changed output by means of a change of inputs (doing so does not consider in the profit function the advantages of an increased mark-up for innovators), plus the advantage of the augmented productivity that might be measured by a fraction of the difference between the current level of flexible inputs and the levels of flexible inputs necessary to produce the desired level of output relying on the same technology. Hence (1.7) can be written:

$$W = (M(Y_1 - Y_0) + F(I_1 - I_0))/Y_0 \qquad (1.7)$$

where W expresses the unit revenue derived from innovating and F is the fraction of increased TFP that the innovating firm can retain. On the basis of (1.6) and (1.7) gross revenue can be expressed so that the total advantage derived from the introduction of technological innovation and the increased level of inputs is:

$$R = W(TC) + Z(S) \qquad (1.8)$$

With high levels of appropriability and the advantage of barriers to imitation, innovation is likely to generate higher than usual mark-ups. Hence, the iso-revenue curve is flatter, the larger are the appropriability regimes and innovation mark-ups.

The rate of substitution between the resources necessary to employ a technique consistent with the previous dimension of flexible inputs and the resources that are necessary to change the technology can be expressed by the slope of the frontier of possible changes curve (FPC):

$$S = c(TC) \qquad (1.9)$$

where S measures the extent of input-augmenting-cum-switching activity and TC the extent of innovating activity. Here again we shall assume that $dS/dTC < 0$ and $dS^2/dTC^2 > 0$. The slope of FPC expresses the ratio of the marginal productivity of flexible-input-augmenting-cum-switching activities to the marginal productivity of innovation activity. The position of the FPC depends on the total amount of resources that can be devoted to cope with the changes that are required by the shift in the demand curve for the firm when attempting to minimize the amount of adjustment activity to introduce technical change for either switching or innovating.

The firm that maximizes profits, net of adjustment activities, chooses whether to adjust its levels of inputs to meet the changed levels of demand or to retain its current level of inputs, changing technology and thus adjusting its output to the new required levels of demand by means of an expanded efficiency of its production factors, according to the relative productivity of the flexible inputs necessary for changing the current levels of output for a given technology, with respect to the productivity of research and development activities necessary for introducing technological innovations. With high levels of localized learning, important barriers to imitation and entry that protect the mark-up levels and strong appropriability regimes, firms coping with rapidly changing demand dynamics are likely to change the levels of outputs more than their input levels.

Formally, a firm will choose whether to change the level of flexible inputs to adjust output or to change technology while retaining the current levels of flexible inputs according to the results of the constrained minimization of adjustment activities. The standard procedure thus leads to optimizing the net profit equation by setting the partial derivatives of the revenue equation and the frontier of possible adjustments equal to zero so that the levels of adjustment activities are minimized where:

$$\mathrm{d}S/\mathrm{d}(TC) = -W/Z \qquad (1.10)$$

From (1.10) it is clear that the value of TC (i.e. the equilibrium level of innovative activity) depends on the marginal productivity of innovation activities and switching activities. The firm will cope with demand growth by varying the combinations of innovation and switching. More precisely:

1. The stronger the increasing returns, the less likely it is that the pressure of demand will induce the introduction of better technology.
2. The lower the value of a in (1.2), the higher are switching and average costs after the increase of flexible inputs (given super-fixed inputs), and the more likely it is that demand pressures will induce the adoption of better technology.
3. The larger the value of b in (1.5), i.e. the greater the efficiency of new production after the introduction of new technology, the more likely are demand pressures to induce the introduction of better technology.
4. The higher the value of b in (1.5), i.e. the greater the learning opportunities that reduce the overall cost of introducing better technology, the more likely it is that demand pressure will lead to the introduction of better technology. Hence, the greater the technological opportunities, the larger are the expected effects of the growth in demand on productivity growth.

5. The smaller the value of F in (1.7), i.e. the better the appropriability conditions and the larger the scope for their application, the larger are the opportunities for firms to retain a larger share of the quasi-rents associated with the introduction of technology.

6. Also, the larger the average mark-up associated with innovation, the greater the innovations, and the larger are the chances of a positive relationship between demand growth and technological change.

7. The longer the expected duration of the quasi-rents stemming from technological innovation, the larger is the expected flow of innovations, and the stronger the causal relationship between growth in output and rate of introduction of technological innovation.

8. The faster the rate of growth of demand, the larger the incentive for firms to search for new technology.

A corner solution is reached when the levels of switching costs are high, the levels of innovation costs are low, profits engendered by innovation are high and profits engendered by the sheer increase in variable inputs are low. This outcome implies that firms will increase their inputs only with the introduction of localized technological change. This means that firms will experience a localized growth of TFP and hence an actual reduction of average costs even below the minimum levels obtained with perfect flexibility of all production factors. Facing these conditions the firm will move rightward, increasing the size of output, along a trajectory defined by the irreversible levels of the super-fixed productive factors. Demand pressure, under these conditions, will push both the rate and the direction of technological change. Firms, facing a continual growth of demand and output, will be induced to generate a stream of technology changes along a technological trajectory by the complementarity of new technology with existing super-fixed productive factors.

Now assume that the levels of the super-fixed production factor can change smoothly in the long term, but only slowly. The growth path of the firm, especially when it relies heavily on the introduction of localized technological change in order to cope with an increase in demand levels, now follows a path dictated by the increasing availability of the super-fixed productive factors. This model provides a microeconomic foundation for understanding the path-dependent cumulative introduction of technical innovations, characterized by dynamic economies of scope based on the complementarity with irreversible super-fixed productive factors. With given super-fixed productive factors firms try to minimize the increase of flexible inputs. It is clear the super-fixed production-factor intensity increases because of the introduction of localized technological innovation. It is worth recalling that the new technological systems will be centred on super-fixed productive factors, which can both be material and immaterial. In the former case the new emerging technological systems will be shaped by the characteristics of the capital structure in place. This is clearly the case for advanced telecommunications. In the latter case the new technological system will be shaped by the competencies accumulated by firms. This model provides a theoretical background for the resource-based theory of the firm (Penrose, 1959).

This approach sheds light on the efficiency of market selection of rival technology. The market selection of new technology is based on comparisons between the average total cost of firms using rival technology: firms and technologies with higher average costs are sorted out through the competitive process. This process implies that the selection of new localized technology is heavily influenced by the pre-existing capital structure of incumbents. Specifically, a technology can appear less efficient than those that rely more on the existing super-fixed capital. In this context the market selection of technology is likely to be affected by the (perverse) effects of dynamic 'lock-ins' which reproduce themselves over time because of the long-lasting effects of super-fixed and sunk productive factors. The innovative efforts of firms and the outcome of the market selection are heavily influenced by the distribution among firms of specific levels of super-fixed productive factors and related sunk costs, together with the combination of market structures and competitive advantage. Finally, the model provides a framework to understand, and possibly implement, the technology strategy of firms characterized by important super-fixed productive factors. For such firms technological strategy can yield a competitive advantage when it mobilizes all possibilities to introduce technology in such a way as to enhance the capability of super-fixed productive factors to obtain potential economies of scope and density. Economies of scope and density require additional and complementary innovation to be fully expressed in the market place.

APPENDIX B: MAPPING THE DIRECTION OF TECHNOLOGICAL CHANGE

Technological change in telecommunications markets can be classified according to its principal economic effect on the conduct of both incumbents and innovators. Technology is usefully classified as either centripetal or centrifugal technology, and these clusters are analysed in turn.[4] Centripetal technology enhances economies of scale, scope and density, and values complementarity with existing capital stock by the incumbent fixed network operators.

1. Coaxial cable systems were first introduced in 1946. They consisted of four pairs of coaxial cables and had a total capacity of 1800 two-way voice circuits. Coaxial cable technology considerably reduced transmission costs and had important effects on increasing returns, in terms of both economies of density and scope for switching activity. The capacity of switching centres is better exploited by new transmission capacity of cables.
2. L4 transistor systems in repeaters were introduced in 1967 and so increased the transmission capacity of coaxial cables (they carried 32 400 two-way voice circuits over 11 pairs of coaxial). The last generation of repeaters, L5, introduced in 1978, enhanced the transmission capacity to 132 000 two-way voice circuits and used integrated circuits. The centripetal effects were stronger than those exerted by L4 technology.
3. Digital switching was introduced in 1968. Digital switching allowed the sharing of the transmission capacity with many different users, converting individual signals to a digital format and then combining the digital signals in time (time-division multiplexing) adding on to the centralizing role of the switching system.
4. Optical fibre, introduced in the late 1970s, was a strand of optical fibre which carried two gigabits (equivalent to 30000 telephone circuits). Each fibre transmission system had dozens of fibre strands in each cable. Plummeting costs and the enormous increase in capacity meant that there were substantial gains in economies of scale. Once a network was established, the incentive to increase traffic volumes was strong. Average costs continued to fall, and marginal costs were lower than average costs.
5. The intelligent network, introduced in the late 1980s, is considered a radical innovation in signalling which, together with switching, transmission, distribution and billing, is one of the most relevant functions within a network. It deeply affects the features of the production process

and leads to important economies of scale, density and scope. Intelligent networks consist of compact and inseparable systems of advanced computers and software capacity and perform distinct yet interdependent functions. They shape the flow of the traffic within the network. At the same time they keep track of the billing, and provide a variety of advanced services such as the 'follow me' or 'voice messages'. The intelligent network is capable of providing centrally the whole array of services available on virtual and local networks by private branch exchange equipment located on the customers' premises. It thus enables centralized network operators to 'strike back' at specialized and virtual networks, offering the same range of new innovative services throughout the network at much lower costs. Because it relies on strong knowhow and software content the average costs of the intelligent network are affected by advanced sunk costs. Hence the larger the numbers of systems delivered, the lower are costs. Consequently, the operators of large networks have relatively lower costs.

6. Digital time division switching was introduced in the early 1980s, and enhanced switching capability, thus increasing capacity and speed and enabling the use of both voice and data communication.

7. Integrated Services Data Network (ISDN) and Broadband Integrated Service Data Network (BISDN), also introduced in the late 1980s, are integrated multi-service and multi-purpose networks that provide a wide range of services including voice, data and images. Within the framework provided by Asynchronous Transfer Mode (ATM) switching technology and intelligent network signalling systems, these systems make it possible to reconcile within a large, unified, centrally managed and planned network the provision of different services to different customer groups.

8. ATM, introduced in the early 1990s, is a new switching and multiplexing technology designed for broadband multi-service telecommunications. It allows the compression of signals, increasing the transmission capability of the existing cable infrastructure.

9. Digital Extended Cordless Telephony (DECT), introduced in the early 1990s, is a cellular telecommunication technology and relies on microwave cells; it is closely linked to the fixed network and operates with a limited range of autonomy. As such, DECT seems designed to maximize the potential economies of scope of the large communication capacity of the fixed network.

10. Asymmetrical Digital Subscriber Lines, introduced in the mid-1990s, allow multiple, simultaneous high-speed services to be carried over existing twisted copper wire pairs, thus substantially increasing the productive potential of the installed copper network.

11. Interactive video communication, introduced in the mid-1980s, allows video communication and interactive TV. This development encouraged the merging of broadcasting and telecommunications. In fact this is considered a complementary technology to optical fibres. Because of the huge transmission capability, in terms of both capacity and speed, optical fibres can carry a wide array of communication services, including data communication and images. Enormous progress in optical fibre technology has brought together the television and the telecommunication industries. The type of traffic carried by optical fibres (hundreds of high-speed television channels and thousands of high speed telephone lines) blurs the distinction between telecommunications and television.

12. Moving Pictures Expert Group, introduced in the early 1990s, is a signal compression technology that makes it possible to reduce the amount of information to be transmitted for multimedia communication by more than a hundredfold.

13. Synchronous Digital Hierarchy also introduced in the mid-1990s optimizes the flow of communication within existing networks.

14. Service Control Point, also introduced in the mid-1990s, consists of specialized modules which are added to the network to introduce new services and reduce the amount of information processed by switches.

15. Wavelength Division Multiplexing technology, introduced in the mid-1990s, makes it possible to transmit simultaneously streams of light of different wavelengths (hence different colours), which can be received separately. This technology greatly reduces the need for electrical amplification in fibre link networks.

16. TCP/IP is a transmission standard which makes compatibility among specific operating standards possible; the Internet relies on TCP/IP.

17. VOI, the new 'voice on the Internet' technology, optimizes packet switching so as to make voice communication on the Internet backbones possible, albeit with some reduction in quality. Its diffusion might favour the full integration of all communication services into a single multi-purpose network.

18. UMTS is a technology that allows users, even in densely populated areas, to rely on mobile telecommunications systems for the full range of communication needs including voice, data and images.

Centrifugal technologies reduce the relevance of technical economies of scale and density. They include specialist technology (which reduces the role of inter-functional telecommunications economies of scope) and segmental technology (which reduces the role of network externalities). Centrifugal technologies were mainly due to mobile telephone operators.

1. Computer-to-computer communications were introduced in the early 1960s, initiated by IBM to enable the remote maintenance of large computers. In the early stages, data communications were mainly used to link plants in order to streamline production and coordinate large bureaucracies. Network externalities were not important while firms' derived demand for telecommunications services multiplied. This segment of demand had a low price elasticity because the opportunity cost of not using the service was technologically unacceptable, and the effects of advanced telecommunications on the competitiveness of firms.

2. Microwave radio systems were first introduced in 1950. Microwave systems drastically reduced the levels of fixed investment, sunk costs and economies of density. Microwave systems are the technology that enabled the entry of Microwave Communications Inc. (MCI) on the St Louis–Chicago route in 1972.

3. Advanced Customer Premises Equipment, introduced in the early 1970s, made it possible to locate large portions of the intelligent capacity of switching services on customers' premises so as to segment the network and enable specialized access.

4. Centrex, introduced in the early 1980s, modularized switching centres, so making it possible to customize and dedicate switching capacity to large users, by-passing the general network.

5. Geo-stationary satellites, introduced in the mid-1970s, had a substantial cost advantage over coaxial cables because of larger capability in terms of volume, with lower fixed costs. Further, geo-stationary technology made it possible to reduce the effect of economies of scale and density in transmission. Finally, in low-density regions, satellites provided an opportunity to by-pass the terrestrial infrastructure. Geo-stationary satellites are now used mainly for long-distance broadcasting.

6. Radio-mobile telephony is a wireless telecommunication system used extensively for mobile communications, which divides a geographical region into cells, uses low-power transmitters within each cell and reuses transmission frequencies in cells that are not contiguous. Radio-mobile telephony requires a new network of cells to be built and so reduces the economies of scope for existing infrastructure. The rapid growth of radio-mobile technology, especially in terms of more efficient uses of frequencies, has created concrete possibilities of replacing the fixed network.

7. Analogue cellular telephony introduced in 1981 could work at 450 MHz, allowing a much better use of Hertzian space.

8. The Total Access Communication System was introduced in the early

1980s to expand the transmission capability of the mobile network into the 900 MHz range.

9. The Global System for Mobile Communication is a cellular mobile technology introduced in 1986 to enhance the communication capacity of Hertzian space by means of the numeric elaboration of signals.

10. The Global Positioning System makes it possible to use satellites to interact with a single mobile unit and coordinate the transmission and switching functions from space.

11. Time Division Multiple Access is an interface technology introduced in the late 1980s to improve the transmission capability of cellular telephony and extend the communication capability of Hertzian space.

12. Mobile Application Part is a protocol elaborated to manage the location of mobile customers in a territory and direct the appropriate transmission signals from satellites.

13. Low Orbit Satellites (LEOs) were introduced in the mid-1990s. LEOs are satellites designed for fixed-point connections that are especially suited for communication with mobile telephones.

14. Personal Communication Networks (PCN) were introduced in the mid-1990s to provide direct wireless access by means of low-power cells analogous to those used in cellular telephony. Their effect, however, is mixed. They by-pass the terrestrial distribution system so that they can be thought of as a new generation of advanced customer-premises equipment; on the other hand, they reduce the size of the cells, so enhancing economies of scope and density with existing terrestrial switching and transmission infrastructure. PCN technology makes it possible to compress further the communication requirements in Hertzian space to 1800 MHz.

15. The Advanced Digital Television System (ADTS), introduced in the mid-1990s, is the new digital high definition television system fully compatible with data communications systems centred on computers and digital broadcasting satellite transmission. The advent of ADTS is an important step toward the implementation of the wireless telecommunication system, and is a direct antagonist of the use of multipurpose cables.

16. Intelligent Antennas, introduced in the mid-1990s, are complex antennas made up of hundreds of cells each of which focus the radio beam and follow it as it moves, so reducing the emitting power and battery consumption and decreasing interference problems.

17. Power-Line was still under experimentation at the end of the 1990s. The eventual success of this technology might allow the transmission of telecommunication services on existing electric cables with scope for entry of competitors from the electric power industry.

NOTES

1. The funding of the European Project Leonardo da Vinci 'L'encyclopedie systemique de la technologie' is acknowledged together with the comments of Jacques De Bandt.
2. The criteria proposed in Helpman (1998) and emphasized by David and Wright (1999) apply.
3. See Appendix B.
4. This mapping is the result of an ongoing implementation effort (see Antonelli, 1999).

BIBLIOGRAPHY

Antonelli, C. (ed.) (1988), *New Information Technology and Industrial Change: The Italian Evidence*, Dordrecht, Kluwer Academic Publishers.

Antonelli, C. (1993), 'The dynamics of technological interrelatedness. The case of information and communication technologies', in D. Foray and C. Freeman (eds), *Technology and the Wealth of Nations*, London, Francis Pinter.

Antonelli, C. (1995), *The Economics of Localized Technological Change and Industrial Dynamics*, Boston, Kluwer Academic Press.

Antonelli, C. (1999), *The Microdynamics of Technological Change*, London, Routledge.

Arthur, B. (1989), 'Competing technologies increasing returns and lock-in by small historical events', *Economic Journal*, 99, 116–31.

Arthur, B. (1994), *Increasing Returns and Path-dependence in the Economy*, Ann Arbor, University of Michigan Press.

Balboni, G.P., Billotti, M. and Saracco, R. (1997), *Paving the Way to the Information Society: The Research Contribution*, Torino, CSELT.

Bijker, W.E. (ed.) (1987), *The Social Construction of Technological Systems*, Cambridge, MA, MIT Press.

Boyer, R., Chavance, B. and Godard, O. (eds) (1991), *Les Figures de l'Irreversibilité en Economie*, Paris, Editions de l'Ecole de Hautes Etudes en Sciences Sociales.

Bresnahan, T. and Trajtenberg, M. (1995), 'General purpose technologies: Engines of growth', *Journal of Econometrics*, 65, 83–108.

Brock, G.W. (1994), *The US Telecommunication Policy Process*, Cambridge, MA, Harvard University Press.

Brousseau, E., Petit, P. and Phan, D. (eds) (1996*)*, *Mutations des Telecommunications des Industries et des Marches*, Paris, Economica.

Callon, M. (1991), 'Reseaux technico-économiques et irreversibilité', in R. Boyer, B. Chavance and O. Godard (eds*)* *Les Figures de l'Irreversibilité en Economie*, Paris, Editions de l'Ecole de Hautes Etudes en Sciences Sociales.

Carlsson, B. (ed.) (1995), *Technological Systems and Economic Performance: The Case of Factory Automation*, Boston, Kluwer Academic Press.

Carlsson, B. and Eliasson, G. (1994), 'The nature and importance of economic competence', *Industrial and Corporate Change*, 3, 687–712.

Carlsson, B. and Stankiewitz, R. (1991), 'On the nature, function and composition of technological systems', *Journal of Evolutionary Economics*, 1, 93–118.

Clark, K.M. (1985), 'The interaction of design hierarchies and market concepts in technological evolution', *Research Policy*, 14, 235–51.

David, P.A. (1975), *Technical Choice Innovation and Economic Growth*, Cambridge, Cambridge University Press.

David, P.A. (1992), 'Heroes, herds and hysteresis in technological history', *Industrial and Corporate Change*, 1, 129–79.

David, P.A. (1993), 'Knowledge property and the system dynamics of technological change', *Proceedings of the World Bank Annual Conference on Development Economics*, Washington, DC, The World Bank.

David, P.A. (1994), 'Positive feed-backs and research productivity in science: Reopening another black box', in O. Granstrand (ed.) *Economics and Technology*, Amsterdam, Elsevier.

David, P.A. and Steinmueller, E. (1994), 'Economics of compatibility standards and competition in telecommunication networks', *Information Economics and Policy*, 6, 217–42.

David, P.A. and Wright, G. (1999), *General Purpose Technologies and Surges in Productivity: Historical Reflections on the Future of the ICT Revolution*, Discussion Papers in Economic and Social History 31, University of Oxford.

Dixit, A. (1992), 'Investment and hysteresis', *Journal of Economic Perspectives*, 6, 107–32.

Farrell, M.J. (1957), 'The measurement of productive efficiency', *Journal of the Royal Statistical Society*, Series A, 120, 253–81.

Fransman, M. (1994a), 'AT&T, BT and NTT: A comparison of vision, strategy, and competence', *Telecommunications Policy*, 18, 137–53.

Fransman, M. (1994b), 'AT&T, BT and NTT: The role of R&D', *Telecommunications Policy*, 18, 295–305.

Helpman, E. (ed.) (1998), *General Purpose Technologies and Economic Growth*, Cambridge, MA, MIT Press.

Henderson, R. and Clark, K.B. (1990), 'Architectural innovation: The reconfiguration of existing product technologies and the failure of established firms', *Administrative Sciences Quarterly*, 35, 9–30.

Latour, B. (1987), *Science in Action. How to Follow Scientists and Engineers in Society*, Cambridge, MA, Harvard University Press.

Misa, T.J. (1995), 'Retrieving sociotechnical change from technological determinism', in M.R. Smith and L. Marx (eds) *Does Technology Drive History? The Dilemma of Technological Determinism*, Cambridge, MA, MIT Press.

Nelson, R.R. (1998), 'The co-evolution of technology industrial structure, and supporting institutions', in G. Dosi, D. Teece and J. Chytry (eds) *Technology Organization and Competitiveness: Perspectives on Industrial and Corporate Change*, Oxford, Oxford University Press.

Owen, B.M. (1999), *The Internet Challenge to Television*, Cambridge, MA, Harvard University Press.

Penrose, E. (1959), *The Theory of the Growth of the Firm*, Oxford, Basil Blackwell.

Perrin, J. (1991), 'Analyse des systèmes techniques', in R. Boyer, B. Chavance and O. Godard (eds) *Les Figures de l'Irreversibilité en Economie*, Paris, Editions de l'Ecole de Hautes Etudes en Sciences Sociales.

Preissl, B. (1995), 'Strategic use of communication technology: diffusion processes in networks and environments', *Information Economics and Policy*, 7, 75–100.

Smith, M.R. and Marx, L. (ed.) (1995), *Does Technology Drive History? The Dilemma of Technological Determinism*, Cambridge, MA, MIT Press.

Soete, L. (1987), 'The newly emerging information technology sector', in C. Freeman and L. Soete (eds) *Technological Change and Full Employment*, Oxford, Basil Blackwell.

Utterback, J.M. (1994), *Mastering the Dynamics of Innovation*, Boston, Harvard Business School Press.

2. Economics of the Internet

Michael D. Pelcovits and Vinton G. Cerf

INTRODUCTION

As of late 2000 the Internet is continuing its rapid growth and evolution. The Internet had its origins as a US Defense Department project to link computers at a handful of research institutions, and has since evolved into a dynamic telecommunications and information medium, which links several hundred million users, several hundred thousand networks, and over 100 million servers worldwide. The aim in this chapter is to provide an insight into the technological and economic forces that fostered the success of the Internet, with the goal of giving a better understanding of how these forces will guide the Internet into the future.

This chapter includes a brief introduction defining the Internet, and a description of what distinguishes the Internet from earlier telecommunications and information networks. The history of the Internet emphasizes how it evolved in an open and flexible manner, guided by a combination of government and marketplace forces. Information on the key institutional features of the Internet, major companies participating in the Internet market and the types of usage of the Internet are given. This provides the background necessary to discuss the economic forces that shape the Internet and to explain how the economics of the Internet are different from those of traditional telecommunications networks. Finally, the chapter concludes with a bit of stargazing as to the future of the Internet.

WHAT IS THE INTERNET?

The Internet is an interconnected network of networks that carries bits of information between computers or smaller networks of computers. The Internet is described as a mesh, because it connects these tens of millions of independent computers, communications entities and information systems, without any hierarchy or rigid controls limiting the paths that users can set up to communicate with each other. What makes this interconnection possible is the use of a set of communications standards, procedures, and formats used by all of the devices connected to the network. The rules of the road, or protocols, that govern the Internet are the Transmission Control Protocol/Internet Protocol (TCP/IP). In fact there are now hundreds of protocols associated with various functions and services of the Internet, but the original basic protocols are TCP and IP, and the collection is sometimes called the TCP/IP suite.

A formal definition of the Internet was approved by the Federal Networking Council in 1995. According to their definition the Internet refers to the global information system that:

(a)   is logically linked together by a globally unique address space based on the IP or its subsequent extensions/follow-ons;

(b)   is able to support communications using the TCP/IP suite or its subsequent extensions/follow-ons, and/or other IP-compatible protocols; and

(c)   provides, uses or makes accessible, either publicly or privately, high level services layered on the communications and related infrastructure.

This definition highlights key aspects of the Internet. A unique addressing scheme that allows users to send messages to the correct destination; an agreed format that allows computers to communicate using a common language; and a flexible platform that can be used to support myriad information services. The Internet is both a network (of networks) and an architecture that provides for these communications capabilities and information processing and exchange. The Internet works by means of protocols that allow the communications to take place along with a layering of user-defined applications, which can be communicated using these protocols. In many ways, this definition supports the characterization of the Internet as an information superhighway, with the same requisite infrastructure and conventions, for example, signs and regulations, used by the interstate highway system to transport all sorts of commodities and vehicles.

In fact it is vital that the notion of Internet not be entirely subsumed in the physical manifestation of networks and computers. The conventions used for access to, exchange of, and processing of information in the Internet are at least as important, if not more important, than any particular physical component. Physical networks come and go, but the Internet persists and evolves. These information conventions create a virtual environment in which digitally encoded information objects are effectively first-class citizens. They interact with software and with one another, they persist and they, too, can evolve. The conventions governing their operation can be thought of as a kind of ecosystem for information. This rather abstract view may prove to be the most important aspect of the Internet as it continues to grow and adapt to new demands.

HISTORY OF THE INTERNET

The earliest incarnation of the primary concepts from which the Internet evolved originated with the US Defense Advanced Research Projects Agency. In the height of the cold war, the US funded research designed to link together computers at universities and other research institutions in the US and in Europe. The name of the new wide-area computer network, taken from the name of the agency funding the project, was ARPANET. This computer network used a new technique for communication, called packet switching, which was far superior for computer-to-computer communication than the technique used for voice telephony, called circuit switching.[1]

The packet switching concept was developed independently by Leonard Kleinrock then of the Massachusetts Institute of Technology, Paul Baran then of the RAND Corporation and Donald Davies of the National Physical Laboratory in the United Kingdom (UK). The basic concept of packet switching is to communicate by sending small bundles or packets of information from one computer to another. In contrast to circuit switching, where a dedicated communication path is established between the two parties for the entire length of a conversation, the individual information packets can each be sent independently, and even on separate routes, so long as they are delivered to the right destination. The communications paths are effectively shared by all the flows of packets rather than dedicated to any one flow. Kleinrock's contribution was to analyze mathematically the performance of communication systems that used finite length messages to communicate – rather like electronic postcards in their character. Baran's work on message block switching – a precursor to packet switching – was developed to support a voice communication system between military command and control centers that would be capable of surviving disruption to the telephone infrastructure during a nuclear war. Davies and his colleagues conceived of packet switching (and coined the phrase 'packet') as a means of facilitating communication between computers in the UK. They actually built one packet switch. In some ways, this might be considered the first local area network, although Davies in fact wanted to construct a national scale system, but was unable to convince the UK government of the value of such an undertaking.

ARPANET was the first wide-area packet-switched computer network.

It successfully deployed packet switches connected by dedicated telecommunication circuits that initially linked together computers in different academic and research centers. The researchers at the various linked institutions created the protocols, that is, conventions, formats and procedures, necessary to allow the computers to exchange information. ARPANET's success showed that developing common protocols for many diverse computers and operating systems, and linking them through a common packet-switched network, enabled computer resource sharing in a way never before accomplished. This success led to the further exploration of packet switching in media other than dedicated communication circuits. In particular, ARPA sponsored the development of two radio-based systems: the ALOHAnet at the University of Hawaii and the Packet Radio Network in a test bed in the San Francisco Bay Area. In addition, ARPA sponsored the development of a trans-Atlantic Packet Satellite network using the synchronous INTELSAT satellites from multiple ground stations. Each of the networks stood alone with essentially no interaction between them.

When ARPA set out to connect a satellite-based network to ARPANET, it initially conceived of linking the networks together by embedding the satellite network's software in the ARPANET packet switches and internetworking the networks through memory-to-memory transfers within the packet switch. In effect, the satellite link would be simply another link in the ARPANET, with the addition of some special software in the packet switches to handle the long-delay procedures necessary to accommodate the 250 millisecond round trip to the satellite in synchronous orbit. Had this course been followed, it would have had only limited benefits and applicability to any later efforts to interconnect additional networks. The evolution of internetworking required a leap forward in architectural thinking, before it could expand its reach and scope to support many networks and many services. Robert Kahn, then at ARPA, developed the idea of an open architecture for internetworking and, together with Vinton Cerf, then at Stanford University, developed the concept of the Internet and its primary architecture and protocols. The creation of the Internet is the story of the evolution from a single packet network, the ARPANET, to a handful of networks (ground mobile packet radio, packet satellite and ARPANET), to the hundreds of thousands of interconnected networks that make up the current Internet.

The Internet evolved from a concept to a reality over a ten-year period, from 1973 to 1983. The Internet architecture needed to be based on a protocol that allowed diverse host computers on diverse networks, of varying degrees of reliability, to communicate with each other. To accomplish this objective, packet-switched communication protocols had to be taken to a new stage of sophistication that would build more intelligence and reliability into the processing of the packets. Furthermore, the protocol had to allow the different component networks to function independently, with few constraints on their own systems for information processing. The Internet protocol suite – TCP/IP – was developed by a large group of collaborating scientists led by Kahn and Cerf. During the first five years of the Internet's development, the TCP/IP protocols were developed. During the later five years, the protocols were implemented on about 30 different computer operating systems and, at the beginning of 1983, the ARPANET was converted to the new Internet protocols, which allowed it to be interconnected with other networks. Among these networks were packet radio networks, packet satellite networks, and local area networks such as the Ethernet, which had been invented by Robert Metcalfe in 1973 while at Xerox Palo Alto Research Center.

The IP piece of the suite supplies the globally unique addresses given to the packets being transmitted among the networks. When the packets are compared to postcards sent on the postal system, then the IP protocol supplies from and to addresses on the postcard. The IP protocol also specifies the digital equivalents of shape, size and other parameters of each postcard. Each of the gateways (the switches or routers on the Internet) is able to read the destination address on the packet, and based on the information and intelligence in the gateway is able to send the packet on the next leg of its journey to its destination. A major reason for the success of the Internet is that its postal system (the gateways and transmission paths) handles all of the routing and delivery of packets, freeing the host computer networks from having to learn or process any information relating to routing. The TCP piece of the suite provides the intelligence that enables the Internet to work, in spite of the fallibility of the electronic postal system that carries the packets. Proceeding with the post office analogy, consider the task of sending a novel by cutting and pasting each paragraph or page of the novel onto a separate postcard, and then dumping the postcards in the mail. How will the recipient be able to reassemble the novel? And more to the point: what will happen when postcards are lost in the mail? The solution is an agreed set of procedures that gives the correct sequence to the packets, resends them if they are thought to be lost, discards any duplicates that are received, and assures that the sender gets confirmation of the receipt of each packet. In addition, TCP exercises control over the rate of data flow between hosts to prevent the network from choking up and the receiver from becoming overloaded. The Internet's success is largely due to the robustness of the protocols it uses to adapt a wide variety of networking and transmission media into a global and coherent architecture. The expression 'IP on everything' is commonly used by Cerf to explain that the

IP is designed to work on any underlying communication substrate. The protocols are designed to adapt to widely varying conditions and capabilities among the component networks and it is this adaptability that has helped the Internet survive and evolve for the last 25 or more years of its existence.

The History of the World Wide Web

The World Wide Web (WWW) was officially conceived in March 1989 in Switzerland by a researcher at what was then the Conseil Europeen pour la Recherche Nucleaire (CERN), the European Laboratory for Particle Physics. Tim Berners-Lee, the father of the WWW, presented the idea for the information sharing system in a paper entitled 'Information Management: A Proposal'. Keeping track of the large quantity of project related data and files that were created by the laboratory was difficult, and resulted in research information being lost. Based on an earlier non-published program that Berners-Lee had written in 1980 for his personal use, the proposal used hypertext to share information among dispersed nodes. Hypertext, the term for which was coined by Ted Nelson in 1965, allowed an author to link a document, any number of times, to other documents. By clicking on the hypertext, the user is transferred to the linked document. The information system proposal also incorporated Hypertext Transfer Protocol (HTTP) as the means of transporting linked documents to the nodes.

Berners-Lee began developing the data management system, which he coined the WWW. He had tacit support but not official support according to his book *Weaving the Web*. The WWW system, which included HTML, HTTP and the client machine software, or browser, WWW, was released to CERN's staff in May 1991. The system was then released free of charge to the High Energy Particle Research community and the public in the summer of 1991. In December 1991, the Stanford Linear Accelerator Center in California installed the first server – the machine on which the documents are stored – outside of Europe. By the end of 1992 there were 50 servers. In February 1993, the National Center for Supercomputing and the University of Illinois released MOSAIC, which was written by Eric Bina and Marc Andreessen.[2] MOSAIC was based on a graphical user interface and allowed people without computer expertise to point and click to navigate the WWW. The program created a surge of public interest in the WWW which has now grown to over 100 million servers.[3]

THE INTERNET AT 2000

Internet traffic initially traverses a link between the customer and an Internet Service Provider (ISP). There are two types of connections between the customer and the ISP: dedicated and dial-up. Dedicated Internet access connects a customer to the Internet via a point-to-point transmission link to the network of the ISP. Dedicated Internet access is offered both on a wholesale and a retail basis, and can range in bandwidth from 128 kbps downstream and 1 Mbps upstream over a digital subscriber loop (DSL) to 1 Gbps in both directions. Access methods range from DSLs to cable modems to shared radio links, such as MMDS, Ricochet services of Metricom and so on. Wholesale dedicated Internet access service is purchased by ISPs that have either limited or no network facilities and that resell Internet connectivity to retail end user customers.[4] Dedicated Internet access is also provided on a retail basis to end users. In the past, the primary customers of retail-dedicated Internet access have been larger businesses. Currently, residential customers are becoming a much more significant source of demand for this service, due to the increasingly widespread deployment of cable modems and DSL services.[5]

In addition to dedicated Internet access, many ISPs offer dial-up Internet access. Customers connect to dial-up access by placing a local telephone call to the ISP's modem banks, which are the portal to the Internet and which convert the analog telephone signal to a digital stream of packets. Like dedicated access, dial-up access is provided on both a wholesale basis and on a retail basis. Wholesale dial-up access service is provided by the larger ISPs to smaller ISPs for resale to end users and involves the provision of access to modem banks as well as a link between the modem bank and the ISP wholesaler's network. The ISP aggregates the dial-up traffic at the modem bank and routes it over a dedicated facility to a node, which, in turn, is interconnected to the Internet.[6] The node contains equipment that recognizes and processes packets and determines the next point in the network to which a packet of information should be forwarded to reach its final destination.

ISPs must connect to other ISPs to form the networks of networks that are recognized as the Internet. These connections can be made using facilities that they own or lease, while others resell services from such facilities-based providers. Facilities-based ISPs are sometimes called Internet backbone providers (IBPs). The facilities necessary to provide backbone service are routers connected together by high-speed data lines. Routers function like switches in the traditional telecommunications network and send individual packets to the correct destination. Inter-city high-speed data lines are almost always provisioned over fiber optic facilities that are available from one of many long-distance carriers. Some of the long-distance carriers are themselves IBPs, but many of the ISPs are not long-distance carriers, but rather lease the long-distance capacity from the long-distance carriers. In addition to the Internet access services described above, many ISPs sell related value-added services, including web hosting, e-mail, co-location, and security products. Web hosting allows a customer to place its web content, for example, a web page, on a server owned and operated by the ISP. Co-location enables large customers to place their own server in space that is owned by the ISP.[7] Security services include the sale and management of firewalls that are designed to permit Internet traffic to flow between the customer's computers and other sites on the Internet, without allowing unauthorized users to access the customer's internal resources.

ISPs may exchange traffic under peering arrangements in which one ISP agrees to deliver Internet traffic to locations on its own network, in exchange for the other ISP's agreement to deliver traffic to locations on its network. Traffic may be exchanged through a public peering arrangement at a network access point (NAP) where multiple peers exchange traffic, or through a private peering arrangement where ISPs establish direct connections for the exchange of traffic. NAPs are, in effect, meeting places where ISPs exchange traffic through individually negotiated peering arrangements. ISPs can lease rack space and cross-connects from the NAP operator in order to link their facilities to other ISPs located at the NAP. The ISPs, not the NAP operator, determine with whom they wish to interconnect. In 1999, there were a total of 44 public peering points in the US operated by numerous firms,[8] an increase of 12 per cent from the 39 NAPs in operation in 1998, and a more than ten-fold increase from 1995.[9] In addition, the recent deployment of high-speed ATM switches at some NAPs has increased the number of ports and the amount of capacity available to ISPs at those locations.

Publicly available information indicates that the demand for and supply of Internet and Internet-related services are continuing to expand at a rapid pace (Kende, 2000).[10] Since 1997 the number of ISPs (facilities-based and resellers) has increased by nearly 40 per cent;[11] the number of points-of-presence per ISP has increased by five times;[12] the number of hosts connected to the Internet has more than quadrupled;[13] and Internet traffic has increased from six to ten times.[14] In addition, there were an estimated 7463 ISPs in the spring of 2000, 20 per cent of which operated on a nationwide basis.[15] Similar to the growth in the number and size of ISPs, the suppliers of Internet connectivity, the IBPs, have grown in number and expanded their physical presence. It is difficult to determine with precision the number of ISPs that operate national backbone networks because different sources may rely on different definitions of what is a backbone service. What is clear,

Table 2.1   Directory of national IBPs

AboveNet Communications, Inc.	Multacom
AT&T	NetRail
BCE Nexxia	One Call
Broadwing	Onyx Networks
Cable & Wireless	OrcoNet
CAIS	PSINet
Cogent Communications	Qwest
Concentric	RMI.NET
Electric Lightwave	Savvis Communications
Epoch	Servint
e.spire	Splitrock/McLeod USA
Excite@home	Sprint
Exodus	Teleglobe
Fiber Network Solutions	Telia Internet Inc.
Genuity Inc.	UUNET
GlobalCenter	Verio
Globix	Vnet
GST Communications	Williams
ICG Communications	Winstar
IDT Corporation	Ziplink
Intermedia	Intira
Level 3	Applied Theory/CRL
Lighting Internet	Road Runner

Source: *Boardwatch* Magazine's Directory of ISPs (2000). Three companies included in this table are not listed in the directory but are discussed in the same publication.

however, is that the number of entities that operate nodes, routers and transmission links that provide access to the Internet is large and is growing. Thus, in its 2000 directory, Boardwatch reported 46 national backbone providers, including nine companies that did not appear in its 1999 directory (see Table 2.1).[16] It is also possible to get some measure of the importance and size of the different backbone providers. Although many measures have been suggested to measure market shares of the IBPs, revenues are used here. An estimate of present and future market shares was recently provided by Sanford Bernstein & Co. It is shown in Table 2.2 overleaf.

USAGE OF THE INTERNET AND THE WWW

The most intensive current use of the Internet is the WWW. This often leads people to be confused about the relationship of the WWW to the Internet.

Table 2.2   Internet backbone revenue and share forecast

Revenue (USD million)	1997	1999	2001	2003
MCI WorldCom	1151	3090	5379	7051
GTE-BBN	346	1207	2375	3860
AT&T	322	924	2206	4120
Sprint	325	728	1148	1660
C&W	233	459	869	1257
All other	287	1677	3326	4186
Total	**2664**	**8085**	**15303**	**22134**
Market share (%)				
MCI WorldCom	43	38	35	32
GTE-BBN	13	15	16	17
AT&T	12	11	14	19
Sprint	12	9	8	7
C&W	9	6	6	6
All other	11	21	22	19
Total	**100**	**100**	**100**	**100**

Source: Hearing on the MCI WorldCom–Sprint Merger Before the Senate Committee on the Judiciary, Exhibit 3, November 4, 1999. Testimony of Tod Jacobs, Senior Telecommunications Analyst, Sanford Bernstein & Co.

The Internet is the global information system that includes communication capabilities and many high level applications. The web is one such application. Other applications include:

(a)   Electronic mail (e-mail), which allows users to send and receive mail and provides access to discussion groups.
(b)   File transfer, which allows a user's computer to rapidly retrieve and save complex files intact from a remote computer. An example of this would be file exchange services, such as Napster.
(c)   Streaming audio and video.
(d)   Telnet, or remote login, which allows a user to log on to another computer and use it as if there.

Virtually every user of the net has access to electronic mail and web browsing capability, and these two applications largely dominate the use of the Internet for most users. The number of e-mail accounts worldwide was more than 569 million at the end of 1999, and is expected to grow soon to a billion accounts.[17]

Size of the WWW

A study of the WWW, released January 2000, by Inktomi and the NEC Research Institute, estimated that based on four months of observation, there were over a billion unique indexable documents on the web.[18] It is important to note that this estimate is only a snapshot of the web at a given point in time. At a growth rate of about 4.5 per cent per month, the web would be significantly larger by the end of 2000 (Dahn, 2000). Furthermore, this figure only refers to indexable pages, there is also a non-indexable part, which by some estimates accounts for about half of the size of the web.[19]

Who is Online

According to eMarketer, an e-commerce research firm, there were 80.8 million Americans online – representing a 30 per cent penetration rate at the end of 1999.[20] The report indicates that in 1999, 58 million of these were active (at least once a week) users and that more than one-half (54 per cent), or 31.3 million, were business users compared with the 26.7 million (46 per cent) who access the web mainly from home. Comparatively, CyberAtlas estimated that there were an estimated 110.3 million US Internet users (including business, educational and home Internet users) at the end of 1999. The US has over five times more Internet users than Japan, the country with the second highest number of Internet users; however, the US share of the Internet is projected to drop to 27 per cent by the end of 2005.[21] Another source, NUA Publish, estimates that as of November 2000, there were 407 million Internet users worldwide, of which 167 million were in the US and Canada.[22] Table 2.3 shows the top ten Internet-using countries at the end of 2000.

Table 2.3   Leading Internet-using countries at the end of 2000

Rank	Nation	Internet users (million)	Share (%)
1	US	135.7	36.2
2	Japan	26.9	7.2
3	Germany	19.7	5.1
4	UK	17.9	4.8
5	PR China	15.8	4.2
6	Canada	15.2	4.1
7	South Korea	14.8	4.0
8	Italy	11.6	3.1
9	Brazil	10.6	2.8
10	France	9.0	2.2

Online Commerce

Business is making increased use of the Internet for purposes of selling to each other, business-to-business (B2B), and selling to consumers, business-to-consumer (B2C). At present, large corporations dominate online commerce. By one estimate, in 1997 the top 10 per cent of e-commerce sites accounted for 90 per cent of online commerce, and 33 per cent of large corporations were conducting business online versus 4 per cent of small businesses.[23] This situation may change in the next few years as small businesses will find it easier to use the Internet and also realize the imperative of using the Internet for dealing with big companies. An April 2000 report by the Yankee Group predicts that US B2B commerce will experience robust growth in the next few years, estimating that 740 billion US dollars (USD) in commerce at the end of 2000 could grow to nearly USD2.78 trillion by 2004. By this time, companies are expected to be purchasing about 30 per cent, and in some industries 50 per cent, of their goods and services online.[24]

Forrester Research estimates even more rapid growth in B2B commerce in the US to USD6.9 trillion in 2004, accounting for 8.6 per cent of all worldwide sales of goods and services.[25] The eBusiness Report, December 1999 from eMarketer indicates that the US generated three-fourths of global e-commerce revenue but estimates that the US portion will fall to about 53 per cent by 2003.[26] Forrester attributes this drop in US influence to the onset of hyper growth in the Western European and Asia-Pacific online economies by 2001 and 2004, respectively. As the e-commerce revolution continues in the US and other developed countries, its effects are still to be felt in most of the world and it is expected to remain this way for some time. Citing Computer Economics research, CyberAtlas states that in 2000, North America, Europe and Asia will make up 94 per cent of online commerce. By 2003, their contribution will fall slightly to 93 per cent.[27] The continued development of e-commerce worldwide, especially in developing countries, will be contingent on business and governments overcoming telecommunications infrastructure issues, unreliable postal services, fear of fraud, complex and costly tariffs, and a host of other obstacles.[28]

The Internet is evolving at such a breakneck pace that the snapshot displayed here will rapidly date. The names and sizes of the players in the industry will change, as will the nature, intensity, and geographic distribution of usage. What is likely to remain constant, however, is the decentralized nature of the Internet. Unlike a conventional telephone network, the Internet is not defined by, nor does it rely on, specific physical connections between users. Rather, as we discussed earlier, the core of the Internet is a set of conventions that govern the exchange of information over a wide range of physical networks. This has major implications for the economic forces at work in the market, and should guide the way government agencies approach regulatory and antitrust issues pertaining to the Internet.

ECONOMICS OF THE INTERNET

This section analyzes the economic forces at work in the Internet industry, with a particular focus on the provision of Internet backbone services. The market for these services has been the subject of intense scrutiny by government agencies, which have been concerned that consolidation would enable dominant firms to exercise market power and harm rivals and consumers.[29] The discussion begins with a brief overview of regulatory and antitrust measures that have been imposed on traditional telephone carriers. Policymakers and academics are deeply grounded in analysis of and experience with the telephone industry, so it is necessary to understand how this history influences their approach to the economics of the Internet. Only then is the role of network externalities in the analysis of the economics of the Internet discussed. This sets the stage for our analysis of market performance and the potential for any firm or group of firms to gain and exercise market power over the Internet backbone.

REGULATION AND ANTITRUST INTERVENTION IN THE TELEPHONE INDUSTRY

Until five years ago, the provision of local telephone service was a monopoly in virtually all markets in the US. In most states, the telephone monopoly was protected from competition by law. In return for the grant of a franchise monopoly, the local exchange carriers (LECs) were subject to a panoply of regulation imposed and enforced by state regulatory commissions and the Federal Communications Commission (FCC). The rates for most telephone services were filed under tariffs, which required approval by the regulatory commissions. As part of this approval process, the regulators would review the LECs' cost of doing business and determine whether their profit rates were reasonable. While price caps or incentive regulation reduced the scrutiny of the LECs' cost in the last few years, regulatory oversight is still an ever-present part of the daily lives of the LECs. The greatest challenges to policymakers and regulators of the telephone industry have come from efforts to open markets to competition. Although local telephone markets were closed to competition for most of the twentieth century, competition occurred in other market segments, including long-distance service, manufacturing of telephone equipment, customer premises equipment, enhanced services (such as voice mail) and wireless service. Much of the need for government intervention in the industry in the past 50 years (including regulation, legislation, and antitrust law) resulted from the need to control anticompetitive behavior by the LECs, who indeed had powerful incentives and weapons to harm their competitors.

The ability of LECs to engage in anticompetitive conduct, in many segments of the market, stems from their control of the local telephone customer. For the most part, competitors have been unable to serve the customer directly, but instead must rely on the LECs to provide an essential input or 'open the door' to allow competitors to access their customers. Long-distance companies, for example, need access to LEC facilities to connect their long-distance networks to local lines serving their customers. Voice mail companies require special connections at the LEC central offices that forward calls from busy or unanswered lines. Carriers offering digital subscriber loop service require collocation and physical connections to their customers' loops at the central office. The LECs' ability to control access to customers is the reason they are termed 'bottleneck monopolist'.

Historically, competitors had no choice but to rely on the bottleneck monopolist, because neither they, nor any other firm, were able to build local facilities to compete with the LECs. This was due in part to legal prohibitions restricting local competition, but also due to the significant economic barriers to entry in local markets. The investments necessary to provide local telephone service are large and require the commitment of capital that cannot be recovered in the event entry is unsuccessful. These are termed sunk costs, and can constitute a substantial protective barrier behind which a monopolist can attack competitors with impunity. The nature of the local exchange bottleneck monopoly, and the barriers to entry that protect the monopoly from competition, are important factors that distinguish traditional common carriers from ISPs. Whether or not this distinction is sufficient to alleviate concerns over dominance in the Internet backbone market is explored below.

Finally, the government has intervened in the industry using two distinct types of policy instruments: conduct regulation and structural remedies. Conduct regulation consists of requirements designed to force the monopolist to accommodate entry and competition, and in essence act against its own self-interest to maximize profits. For example, laws that forced the Bell System to interconnect with competing long-distance companies are a form of conduct regulation. Structural remedies refer to measures that alter the corporate organization of the monopolist, such as the break-up of the Bell System accompanied by line-of-business restrictions on the divested Regional Bell companies. Controls over mergers and acquisitions are another form of structural remedy. The approach the government has taken to competition issues in the Internet is strongly influenced by its experience with these forms of regulation and control. Conduct regulation requires ongoing, resource-intensive involvement by a government agency. Structural remedies are a surgical strike on the competitive nature of an industry. The reluctance of the US government to risk being compelled to impose conduct regulation on the Internet in the future has undoubtedly led it to take a very aggressive stance on structural issues related to the Internet backbone.

ANALYSIS OF COMPETITION IN THE INTERNET BACKBONE MARKET

Earlier analysis of the share and size of the firms participating in the market showed there are over 40 national backbone providers. Barriers to expansion by existing firms or entry by new firms is considered next. Finally, the ability and willingness of customers to switch from one backbone provider to another in response to an attempt by any provider to raise prices or degrade service is examined.

Barriers to Entry or Expansion in the Internet Transport Market

Entry into the supply of dedicated Internet access services can be achieved relatively quickly and at low costs. Fiber capacity can be leased from a variety of sources, and there is no shortage of capacity available to enable smaller networks or new entrants to expand capacity or enter the market.[30] New fiber carriers are laying cables and creating new backbone networks.[31] Suppliers of Internet connectivity, including AT&T, Global Crossing, Qwest Communications, Williams Communications and Level 3 Communications, have grown in number and expanded their physical presence to meet the increase in demand from other ISPs as well as residential and business customers. It is particularly important that these fiber networks can expand circuit capacity manifold at relatively low cost, in one of three ways. First, they can increase the capacity of existing circuits by using more advanced forms of Wave Division Multiplexing (WDM). Second, they can light up fibers that have been laid, but are presently unlit. Third, some carriers have laid extra conduits (empty pipes), alongside the conduits that carry their fiber. It is relatively inexpensive to run new fiber through these empty conduits and add large amounts of new capacity. The other basic inputs into forming an Internet backbone are the switches or routers that send packets to the right destination. These components are available from a variety of third-party suppliers, which are neither owned nor controlled by IBPs.

In markets where the incumbent has a proprietary standard and an entering rival must promote an incompatible alternative standard – as in operating systems for personal computers – the standards can be a formidable barrier to entry. On the other hand, in markets where all rivals use the same public standard, there is no barrier at all. Rather the use of a single standard can support entry and competition from a large number of rivals. The Internet is based on open standards and protocols that are outside the control of any one of the incumbent backbone providers. It is highly unlikely that backbone providers will implement proprietary standards in the future.[32] There are well-established mechanisms for extending Internet standards. A proposed new standard undergoes a period of development and several iterations of review by the Internet community and revision based on experience before it is adopted as a standard and published.[33] This whole standards process takes place under the auspices of the Internet Society, a non-profit body. The process is managed by the Internet Architecture Board and the Internet Engineering Steering Group, and conducted by the Internet Engineering Task Force.

It is widely recognized that sunk costs constitute the most serious barriers to entry. Sunk costs are not large in Internet service provision. Internet companies do not need to own fiber capacity. Rather, transmission capacity – circuits on a fiber-optic network – can be leased, and a network need only install switches and routers in order to serve its customers. Even those networks that do acquire their own fiber capacity sink relatively little when they invest, because fiber capacity can be used for purposes other than the provision of Internet connectivity. Fiber is essentially fungible, and the same physical networks can be used for the transmission of voice, Internet traffic, and data using other protocols. Fiber that will not be needed by an Internet transport supplier can be leased or sold for non-Internet uses. Finally, fiber can be lit at a wide range of speeds, so that carriers can respond flexibly to increasing capacity requirements.

Ability and Willingness of Transport Customers to Switch to Different Suppliers

When customers are tightly bound to suppliers, through high costs of changing suppliers or long-term exclusive contracts, an important barrier to expansion or entry will result. This type of barrier, however, is almost non-existent in the Internet transport market. First, a large percentage of major ISPs and web sites already use multiple suppliers, a practice called multi-homing. These customers have chosen to avoid any limitation on their ability to switch traffic among suppliers even in the very short run. In August 1999, *Boardwatch* magazine reported that 43 per cent of all connections sold by backbones to ISPs were for multi-homing. Second, ISPs and content providers can easily change suppliers of backbone services provided that they are large enough to have portable addresses. Switching costs of any important magnitude do not lock them in. An ISP customer with a T1 line (1.5 Mbps capacity) would incur costs of under USD10 000 in non-recurring fees to change backbone providers. Third, because of competition backbone providers cannot tie up customers by forcing long-term exclusive contracts on them.

Test of Market Power

The test of whether any firm has market power is to ask whether it could raise price by a non-trivial amount without losing enough customers to make that price increase unprofitable. The answer to this question with respect to the Internet backbone market is no. There are many alternative providers that are willing and able to serve that firm's customers. Competing providers have adequate capacity, or can add capacity, to serve additional demand. New firms are entering the market and can pick up customers that are dissatisfied with any of the incumbent firms. Standard setting is not a barrier to entry, but rather a factor promoting cooperation and compatibility among the firms in the industry. Finally, customers are able to switch providers quickly, if not instantaneously, at relatively low cost. Therefore, based on first principles of microeconomics, the Internet transport market satisfies the conditions of a highly and effectively competitive market.

INTERCONNECTION POLICIES AND NETWORK EFFECTS ON THE INTERNET BACKBONE MARKET

There are circumstances when conventional analysis of barriers to expansion and entry does not account for economic forces that can create or preserve a monopoly. For example, local telephone markets were fully opened to competition by the Telecommunications Act of 1996, yet Congress recognized that removal of entry barriers alone would not be sufficient to allow new entrants to compete successfully. An entrant would be totally unsuccessful at obtaining any customers, were it not able to deliver and send calls between its customers and those of the LEC. Thus the Act requires the LECs to interconnect with the new entrants on non-discriminatory, cost-based terms. Are there similar factors at work in the Internet?

Are some carriers more dependent on others, such that a firm, or group of firms acting in concert, could refuse to interconnect on fair and reasonable terms to exercise control and dominate the Internet? Some of the smaller IBPs have argued that the interconnection practices of the larger IBPs exhibit anticompetitive features. They claim that restrictions on peering imposed by the large carriers are unjustified on an economic basis and exist because of the market power of these carriers. At times, smaller IBPs and government agencies have claimed that discriminatory interconnection practices of the larger IBPs are a consequence of network externalities that reinforce the market power of the largest Internet firms.

Interconnection of the Internet Backbone: Peering versus Transit

Internet networks have contracts that govern the terms under which they pay each other for connectivity. Payment takes two distinct forms: payment in dollars for transit and payment in kind, that is, barter or peering. Connectivity arrangements among ISPs encompass a seamless continuum, including ISPs that rely exclusively on payment for transit to achieve connectivity, ISPs that use only peering to achieve connectivity and sell transit service, and everything in between.[34] Under transit, network X connects to network Y with a pipeline of a certain size, and pays network Y for allowing X to reach all Internet destinations. Under transit, network X pays Y to reach not only Y and its customers, but also any other network, such as network Z that connects with Y. This arrangement is similar to the relationship between subscribers and carriers or between long-distance carriers and local carriers in the telephone market.

Under peering, two interconnecting networks agree to the free exchange of traffic that originates and terminates from the transit customers of the networks. For instance, if Y and Z have a peering agreement, they exchange traffic that originates on Z and terminates in Y and the traffic that originates on Y and terminates on Z. Peering arrangements will not provide for the transit of traffic across one carrier to a third carrier. For example, if X were to enter into a peering relationship with Y, then Y would no longer carry X's traffic to network Z. Peering is similar to the co-carrier relationship that adjacent ILECs have maintained for many years, and is also similar to the bill-and-keep arrangements that some ILECs and competitive local exchange carriers (CLECs) have entered in to.[35]

An important policy issue in regard to peering and transit is whether the difference between the arrangements is related to actual differences in the cost of the network arrangements, or whether the unwillingness of large IBPs to peer with small IBPs is a manifestation of the large IBPs' market power. The starting point for an analysis of this issue is to recognize that the cost to each carrier of agreeing to peer is not zero. When any two networks Y and Z enter into a peering agreement, it means that they agree that the cost to network Y of carrying Z's traffic to its destination within its network is roughly the same as the cost to Z of carrying Y's terminating traffic within its own network. These costs have to be roughly equal if the networks peer, but they are not zero. Under the benign theory of the peering decisions of large IBPs, their decision whether to peer or require transit payments is a commercial one, made without anticompetitive intent or effect. Peering is preferred when the cost incurred by the carriers is roughly the same. If not, the networks will use transit. Two networks could enter into bilateral transit agreements, which result in payments that roughly offset each other and are more accurate than the bill-and-keep aspect of peering. They choose not to do so, because the benefit of greater accuracy is offset by the lower transactions cost associated with peering.[36] Alternatively it can be hypothesized that peering practices can be used anticompetitively by larger IBPs to raise the costs of their smaller rivals. When the large IBPs have market power, individually or competitively, then it would be to their advantage to impose above-cost transit charges on their competitors. This strategy could be effective both in the short run at raising retail rates and increasing profits, and in the long run by driving competitors out of business and further consolidating power in the hands of a few firms. Whether the Internet market is susceptible to these actions, now or in the future, depends on the impact of network externalities and is examined below.

It is worth reflecting on the reason why policymakers are so concerned about the peering issue. If the industry were to evolve to the point that interconnection policies were used by a dominant IBP (or group of IBPs) to restrict competition by smaller carriers, policymakers would be put in a difficult predicament. The government could seek to alter structural conditions in the market, for example by bringing antitrust action designed to split up large IBPs. This would be a time-consuming process with an uncertain outcome. Alternatively, the government could seek to control the behavior of the dominant firm(s) by requiring IBPs to interconnect under terms and conditions stipulated by regulators. In light of the tortuous regulatory process in the telephone industry, it is understandable if policymakers are reluctant to descend the slippery slope and be compelled to apply conduct regulation to the Internet.

NETWORK EXTERNALITIES

The Internet like any network exhibits externalities. Network externalities are present when the value of a service to consumers rises as more consumers use it, everything else equal (see Economides, 1996; Farrell and Saloner, 1985; Katz and Shapiro, 1985; Liebowitz and Margolis 1994). In traditional telecommunications networks, the addition of a customer to the network increases the value of the network connection to all other customers, because it increases the number of customers each party can reach. On the Internet, the addition of a user potentially adds to the information that all others can reach; adds to the goods available for sale on the Internet; adds one more customer for e-commerce sellers; and adds to the collection of people who can send and receive e-mail or otherwise interact through the Internet. Thus the addition of an extra computer node increases the value of an Internet connection to each other node or user.

In general, network externalities arise because high sales of one good make complementary goods more valuable. Network externalities are present not only in traditional network markets, such as telecommunications, but also in many other markets. For example, the Windows operating system is more valuable if there are more computers using the system, because then there will be more software written and sold for such computers. Where there are multiple networks engaged in a similar enterprise, there are large social benefits from the interconnection of the networks and the use of common standards. A number of networks have harnessed the power of network externalities by using common standards or by adopting various ownership structures. Examples of interconnected networks of diverse ownership that use a common standard include the telecommunications network, the network of facsimile machines and the Internet. Despite the different ownership structures in these networks, the adoption of common standards has allowed each one of them to reap huge network-wide externalities. For example, users of the global telecommunications network reap the benefits from network externalities, despite the fragmented industry structure. If telecommunications networks were not interconnected, consumers in each network would only be able to communicate with others on the same network.[37] Thus there are strong incentives for every network to interconnect with all other networks so that consumers enjoy the full extent of the wider network externalities.

Market Power in Industries Exhibiting Network Externalities

Network externalities have had a profound effect in some industries, driving markets to a winner-take-all outcome, where one standard or technology dominates over all others. Under some conditions, the network externalities can cause a market to tip and propel a leading technology to the point where competitive technologies are forced out of the market entirely. This may happen even though the competitors are as efficient and attractively priced as the winning technology, because consumers are interested in more than their own consumption choice. As a result of this phenomenon, policymakers have been much more concerned with the structure and performance of network industries, even where traditional barriers to entry do not exist. Certainly in cases when a firm is endowed with bottleneck monopoly power, the existence of network externalities will reinforce its control over the market. As explained earlier in the case of the local telephone market, the existence of network externalities would make entry by firms trying to break through the incumbents' bottleneck monopoly virtually impossible, if the LECs were not required to interconnect with their rivals. This is a manifestation of the network externalities in telephone markets. Even in markets where bottlenecks do not preexist, however, network externalities may enable a firm to create bottlenecks and exercise market power by using proprietary standards.

GOVERNMENT INTERVENTION AND CONCENTRATION IN THE INTERNET BACKBONE MARKET

Both the US Department of Justice (DoJ) and the Commission of the European Communities (EC) alleged that additional concentration in the Internet backbone industry arising from mergers involving WorldCom had a substantial probability of leading to competitive harm. The case presented by these bodies rested on a chain of logic that can be summarized as follows.[38]

(a) It is important to distinguish between the tiers of IBPs. Tier 1 providers typically maintain private peering relationships with all other Tier 1 IBPs. Tier 2 providers typically purchase transit from one or more Tier 1 providers, although they may engage in some amount of peering. The DoJ alleged that lower-tier IBPs that must purchase a significant amount of connectivity from other IBPs operate at substantial cost disadvantages compared to Tier 1 IBPs, which rely exclusively on peering.[39]

(b) The Tier 1 industry, which according to the DoJ constitutes a relevant antitrust market, is highly concentrated. WorldCom/UUNET is the largest provider by far. Further consolidation would change the balance of power in the market and alter the dominant firm's incentive to cooperate with smaller firms on interconnection terms. The presence of network externalities means that the larger network would become much more valuable to the smaller network than the smaller network is to the larger network. This would provide the larger network with the capability to leverage its size and disadvantage its rivals.

(c) The dominant firm would attempt to degrade the quality of interconnection provided to its rivals, in order to induce customers to switch and buy service from itself. This strategy could have a life of its own and lead to tipping of the market or snowballing as the dominant firm would get stronger and its rivals even weaker. The ultimate results could be total monopolization of the Internet backbone by one firm.[40]

The DoJ's and the EC's concerns about the potential for harm resulting from an increase in the size of WorldCom's Internet backbone led the agencies to impose conditions on the WorldCom/MCI merger requiring the divestiture of MCI's Internet assets. Similar concerns also contributed to the expected rejection and eventual dissolution of the proposed WorldCom/Sprint merger. As discussed earlier, the DoJ's reluctance to risk slipping on the slope to regulation, and repeating the experience of telephone regulation, was a very important factor its decision. In the complaint it filed against the WorldCom/Sprint merger, the DoJ stated that if the market tipped to greater dominance, 'restoring the market to a competitive state often requires extraordinary means, including some form of government regulation'.

PRO-COMPETITIVE EFFECTS OF NETWORK EXTERNALITIES ON THE INTERNET BACKBONE MARKET

Contrary to the position taken by the DoJ and the EC, there are many empirical and logical reasons to believe that the Internet backbone market is not susceptible to tipping or snowballing, and that in fact network externalities serve to enhance competition in this business. Economics literature has established that using network externalities to affect market structure by creating a bottleneck requires three conditions: networks use proprietary standards; no customer needs to reach nodes of or to buy services from more than one proprietary network; and customers are captives of the network to which they subscribe and cannot change providers easily and cheaply (see Farrell and Saloner, 1985; Katz and Shapiro, 1985; Economides, 1989 and 1996).

Without proprietary standards, a firm does not have the opportunity to create the bottleneck. When proprietary standards are possible, their development by one network isolates its competitors from network benefits, which then accrue only to one network. The value of each proprietary network is diminished when customers need to buy services from more than one network. The more consumers are captive and cannot easily and economically change providers, the more valuable is the installed base to any proprietary network. The example of snowballing network effects often discussed in the literature – the victory of VHS against Beta – fulfills these conditions.

It is apparent that the Internet does not fulfill any of the necessary conditions under which a network may be able to leverage network externalities and create a bottleneck. First, there are no proprietary standards on the Internet, so that condition fails. The scenario of standards wars has not applied to Internet transport, where full compatibility, interconnection, and interoperability prevail. For Internet transport, there are no proprietary standards. There is no control of any technical standard by backbone providers and none can be foreseen. Internet transport standards are firmly public property (see Kahn and Cerf, 1999; Bradner, 2000). As a result, any provider of backbone service can create a network complying with the Internet standards – thereby expanding the network of interconnected networks – and compete in the market. In fact, the existence and expansion of the Internet and the relative decline of proprietary networks and services, such as Prodigy, can be attributed to the conditions of interoperability and the tremendous network externalities of the Internet.[41] America On Line (AOL), Prodigy, MCI and AT&T folded their proprietary electronic mail and other services into the Internet. Microsoft, thought to be the master of exploiting network externalities, made the error of developing and marketing the proprietary Microsoft Network (MSN). After that product failed to sell, Microsoft re-launched MSN as an ISP, adhering fully to the public Internet standard. This is telling evidence of the power of the Internet standard and demonstrates the low likelihood that any backbone provider can take control of the Internet by imposing its own proprietary standard.

Customers on the Internet demand universal connectivity, so this condition fails. Users of the Internet do not know in advance what Internet site they may want to contact or to whom they might want to send e-mail. Thus Internet users demand from their ISPs, and expect to receive, universal connectivity. This is the same expectation that users of telephones, mail and facsimile machines have: that they can connect to any other user of the network without concern about compatibility, location or, in the case of telephone or facsimile, any concern about the manufacturer of the appliance, the type of connection (wireline or wireless) or the owners of the networks over which the connection is made. Because of the users' demand for universal connectivity, ISPs providing services to end users or to web sites must make arrangements with other networks so that they can exchange traffic with any Internet customer. This does not exclude the possibility, however, that large application providers will attempt to develop proprietary applications supported by layers of protocol well above IP. These providers may be successful in creating uses of the Internet that are incompatible with other applications. Nevertheless, this is highly unlikely to undermine the fundamental open standards that govern the Internet backbone market. Finally, there are no captive customers on the Internet, so the final condition fails, for a number of reasons: ISPs can easily and with low cost migrate all or part of their transport traffic to other network providers; many ISPs already purchase transport from more than one backbone to guard against network failures and for competitive reasons (multi-homing); many large web sites use more than one ISP for their sites (customer multi-homing); and competitive pressure from their customers makes ISPs agile and likely to respond quickly to changes in conditions in the backbone market.

While the Internet is currently structured in a way that network externalities are a pro-competitive force, it is legitimate to ask whether this will always be the case, or whether at some point a large IBP would attempt to move its traffic onto a proprietary network. Shapiro and Varian note that:

> This situation may well change in the future as new Internet technology allows providers to offer differential quality of service for applications such as video conferencing. (Shapiro and Varian, 1998: 46)

Although this is certainly a theoretical possibility, we believe it is highly unlikely to occur without a fundamental change in the structure of the Internet industry and in the interests and incentives of major users of the Internet. Many large and powerful companies rely on the Internet as the lifeblood of their business, for example Amazon, AOL, Yahoo. Other companies, such as CNN or the New York Times, reap substantial benefits from their ability to reach customers using an inexpensive telecommunications and information technology that is not under any other parties' control. These companies will not be idle in the face of an attempt by a large IBP to control access to their customers. This would be analogous to a publisher being forced into selling its magazines or books through a single distribution channel, thereby sacrificing control and profitability. The market forces that drive compatibility on the Internet are powerful and inescapable in the world as we now know it.

FUTURE OF THE INTERNET

Cerf (2000) considers that the Internet will eventually seem to disappear by becoming ubiquitous. Appliances, automobiles, houses, and countless other devices will become Internet enabled, and communicate with each other over high-speed, low-power radio links. The core of the network will remain optical fiber, capable of transmitting trillions of bits per second, while access will rely on a number of technologies, ranging from radio and infrared to fiber and copper telephone and cable lines. These advances will allow objects to communicate and act on the information being exchanged. For example, automobiles will communicate orders for replacement parts, prior to being brought into a shop for repair. Medical testing and procedures will be able to take advantage of information embedded in the human body or maintained in a distant location.

What are the implications of these likely developments for competition? Is it possible for a single firm to mushroom and control key segments of the Internet, as the Bell System did in the telephone market? Or are there likely to be antitrust problems similar to those concerning Microsoft in the late 1990s? These are difficult questions to answer yet we believe the Internet is unlikely to succumb in the future to the control of a single firm or small group of firms. Ownership of the physical network carrying Internet traffic is unlikely to be concentrated in a few hands, unless there are major public policy blunders. Long-distance and big city fiber networks are proliferating. Local access will eventually be provided over a variety of wired and wireless alternatives, barring undue concentration in the ownership of spectrum, or restrictions on access to rights of way. Concentration in the ownership or control over the logical part of the Internet also seems unlikely. The demand for ubiquity, the power in the hands of users of the Internet, and the flexibility in the ways that physical connections can be made, all make it extremely difficult for a single firm to dominate the Internet. In addition, these factors and the complexity of this market make it extremely difficult for a small group of firms to collude and control the Internet. The analogy to Microsoft is also weak, because the standards that enable compatibility of different systems are not captured in proprietary software. To conclude it is likely that the Internet will not require economic regulation or become subject to antitrust scrutiny as it matures. Governments will increasingly be focused on other issues such as privacy, taxation, intellectual property and contracts, and leave the market to sort out the economic issues.

NOTES

1. Circuit switching is still used today for the vast majority of voice telephony but the situation is evolving towards a data-centric target.
2. Andreesen, along with Jim Clark, former founder of Silicon Graphics, formed a company in 1994, later known as Netscape Communications.
3. www.netsizer.com.
4. Dedicated Internet access service is also called transit. See also Application of WorldCom and MCI Communications Corporation for Transfer of Control of MCI Communications Corporation to WorldCom Inc., CC Docket No. 97-211, Memorandum Opinion and Order No. 146. 'In a transit arrangement between ISPs, one ISP pays another in return for the second ISP's agreement to deliver all Internet traffic that originates or terminates on the first ISP's network, regardless of the destination or source of that traffic.'
5. The FCC recently observed that DSL line deployment is projected to increase from 575000 by the end of 1999 to more than 7.6 million by the end of 2002. *Deployment of Wireline Services Offering Advanced Telecommunications Capability*, Third Report and Order, CC DKT. 98-147, FCC 99-355 (rel. 9 December 1999) at n. 8. According to a more recent estimate of residential broadband subscription, the number of DSL customers reached 1.4 million and the number of cable modem customers reached 2.9 million in 2000. See eMarketer, 'The business of broadband', September 2000.
6. If the ISP only operates the modem bank, it provides access to the Internet by purchasing wholesale-dedicated Internet access service from another ISP.
7. Co-location hosting enables a customer to place its own server in space that is owned by the co-location provider. Customers of co-location hosting generally provide their own monitoring and maintenance services. Shared web hosting involves the provision of web-hosting services to multiple customers whose websites are maintained on a single server owned and maintained by the web-hosting provider. Dedicated hosting involves the provision of web-hosting services to a single customer whose website is maintained on one (simple) or several (complex) separate servers owned and operated by the web-hosting provider. Managed hosting involves complex, dedicated hosting arrangements in which the web-hosting provider also performs administration, monitoring and maintenance services.
8. See http://www.ep.net/naps_na.html.
9. See 'The Internet – What is It?', *Boardwatch Magazine*'s Directory of ISPs (11th edn, 1999) at http://boardwatch.internet.com/isp/summer99/internetarch.html. In 1994, the NSF announced that four NAPs would be built.
10. WCOM, Worldcom at the Heart of Internet Growth, 1999 e-Annual Report at http://www.wcom.com/about_the_company/investor_relations/annual_reports/1999/do -internet.phtml; see also Alex Gove, *Kings of the WorldCom*, *Red Herring Magazine*, http://www.redherring.com/mag/issue59/kings.html, 2 October 2000.
11. Cahners In-Stat Group, The US ISP Industry: What is it Earning? What is it Spending?, Table 2, Vendor-Projected US ISP Market Sizing, 1997–1998, Report IS99-01MC at http://www.instat.com/abstracts/ia/1999/is9901mc_abs.htm, April 1999.
12. Ibid., at Figure A4: Average Number of Points of Presence Per ISP, 1997–1998.
13. M. Lottor, Survey, Network Wizards, http://isc.org/ds/WWW-200007/index.html, July 2000. A host is defined as a computer connected to the Internet with a static IP address that can respond to a query.
14. Gilder Technology Group, 1999 Newsletter, at http://gildertech.com/html/gtg.html
15. Michael Robuck, 'Report Says National ISPs will Dominate US Market', ISPworld at http://www.ispworld.com/bs/BS_92600a.htm; Cahners In-Stat Group, 'The US ISP Industry: Revenues and Services,' 21 September 2000, at http://www.instat.com/ abstracts/ia/2000/is0004sp_abs.htm.
16. The nine companies listed in Boardwatch's 2000 directory that did not appear in the 1999

directory are: BCE Nexxia, Cogent Communications, Lighting Internet, Multacom, NetRail, One Call, Onyx Networks, OrcoNet, Applied Theory/CRL.

17. http://www.messagingonline.com/mt/html/feature031400.html.
18. www.inktomi.com/webmap.
19. http://www.infotoday.com/newsbreaks/nb0712-1.htm.
20. http://www.emarketer.com/eservices/101999_profile.html.
21. http://cyberatlas.internet.com/big_picture/geographics/article/0,1323,5911_151151,00. html.
22. http://www.nua.com/surveys/how_many_online/index.html.
23. http://www.iw.com/daily/stats/1998/05/2103-ecommerce.html.
24. http://www.yankeegroup.com/webfolder/yg21a.nsf/press/78BEE5798DFEA62D852568 C800619B77?OpenDocument.
25. http://www.forrester.com/ER/Press/Release/0,1769,281,FF.html.
26. http://www.emarketer.com/estats/sell_ebiz.html.
27. http://cyberatlas.Internet.com/big_picture/geographics/print/0,1323,5911_309941,00. html.
28. 'The e-Global Report', March 2000, eMarketer.
29. Whether the provision of Internet backbone services constitutes a market from an anti-trust standpoint is debatable, and in fact has been a key controversial subject in the reviews of mergers involving WorldCom. For purposes of the discussion in this chapter the term 'market' is used in a non-technical sense, without meaning to endorse the view that the Internet backbone constitutes an antitrust market.
30. See 'In the Matter of Inquiry Concerning the Deployment of Advanced Telecom-munications Capability to All Americans in a Reasonable and Timely Fashion, and Possible Steps to Accelerate such Deployment Pursuant to Section 706 of the Telecommunications Act of 1996', Second Report, CC Docket No. 98-146, FCC 00-290 at No. 208 (rel. 21 August 2000) (Advanced Services Report).
31. IDC, *Emerging Internet Service Providers* (August 2000).
32. This does not preclude the possibility that a company with power in a related market, such as Microsoft, could attempt to leverage that market power and impose standards that would spill over into the Internet.
33. Scott Bradner, 'The Internet Standards Process', revision 3, Network Working Group at ftp://ftp.isi.edu/innotes/rfc2026.txt, p. 4.
34. Transit customers receive services, such as customer support, DNS services and so on, that peering networks do not receive.
35. Recently there has been an increased interest in the use of bill-and-keep arrangements in the telephone industry. LECs have sought to end the reciprocal compensation mecha-nisms, in place for the last several years, which have caused them to pay CLECs hundreds of millions of dollars. Recently an FCC staff working paper proposed mandating bill-and-keep as a default interconnection regime among all telecommunications carriers. See DeGraba (2000).
36. Generally, peering does not imply that networks should have the same size in terms of the numbers of ISPs connected to each network, or in terms of the traffic that each of the two networks generate. When networks are similar in terms of the types of users to whom they sell services, then the amount of traffic flowing across their interconnection point(s) will be roughly the same, irrespective of the relative size of the networks. What determines whether a peering arrangement is efficient for both networks is the cost of carrying the mutual traffic within each network. This cost will depend crucially on a number of factors, including the geographic coverage of the two networks. Even if the types of ISPs of the networks are similar, and therefore the traffic flowing in each direc-tion is the same, the cost of carrying the traffic can be quite different in network X from network Y. For example, network X (with the ten ISPs) may cover a larger geographic area and have significantly higher costs per unit of traffic than network Y. Then network X would not agree to peer with Y. These differences in costs ultimately would determine the decision to peer (barter) or receive a cash payment for transport. Where higher costs

are incurred by one of interconnecting networks because of differences in the geographic coverage of each network, peering would be undesirable from the perspective of the larger network. Similarly, one expects that networks that cover small geographic areas will only peer with each other. Under these assumptions, who peers with whom is a consequence of the extent of a network's geographic coverage, and does not have any particular strategic connotation.

37. By contrast, telex networks were not interconnected, which significantly limited the benefits to the users of this technology.
38. Complaint, in the US District Court for the District of Columbia, USA DoJ versus WorldCom and Sprint Corporation, 26 June 2000, at http://www.usdoj.gov/atr/cases/indx239.htm. The Court proceedings ended when WorldCom and Sprint cancelled their plan to merge. The EC case results are not public documents, but the Commission's reasoning is similar to that of the DoJ.
39. Ibid., par. 28.
40. In the course of the EC proceedings on the WorldCom merger with MCI, a paper by Professors Cremer, Rey and Tirole was presented on behalf of GTE. This paper alleged that the dominant firm would pursue a strategy of serial degradation, whereby only the weakest rival would be targeted. On its demise the dominant firm would then target the next weakest rival, and so on, until all rivals were destroyed.
41. Some networks, such as Compuserve, continued to operate specialized services, such as transaction processing, over systems that were not interconnected with the Internet.

REFERENCES

Bradner, S. (2000), The Internet Standards Process, revision 3, Network Working Group (ftp://ftp.isi.edu/in-notes/rfc2026.txt), Section 1.2.

Cerf, V.G. (2000), 'What will replace the Internet', *Time*, 19 June 102.

Dahn, M. (2000), 'Counting angels on a pinhead: Critically interpreting web size estimates', ONLINE, January.

DeGraba, P. (2000), 'Bill and keep at the central office as the efficient interconnection regime', OPP Working Paper 33, Washington, DC, FCC, December.

Economides, N. (1989), 'Desirability of compatibility in the absence of network externalities', *American Economic Review*, 79(5), 1165–81.

Economides, N. (1996), 'The economics of networks', *International Journal of Industrial Organization*, 14(2), 675–99.

Farrell, J. and Saloner, G. (1985), 'Standardization, compatibility, and innovation', *Rand Journal of Economics*, 16, 70–83.

Kahn, R.E. and Cerf, V.G. (1999), 'What is the Internet (and what makes it work)', at http://www.wcom.com/about_the_company/cerfs_up/internet_history/whatIs.phtml, December.

Katz, M. and Shapiro, C. (1985), 'Network externalities, competition and compatibility', *American Economic Review*, 75(3), 424–40.

Kende, M. (2000), 'The digital handshake: connecting Internet backbones', OPP Working Paper No. 32, FCC, September.

Liebowitz, S.J. and Margolis, S.E. (1994), 'Network externality: An uncommon tragedy', *Journal of Economic Perspectives*, 133–50.

Shapiro, C. and Varian, H.R. (1998), *Information Rules: A Strategic Guide to the Network Economy*, Cambridge: MA, Harvard Business School Press.

3. Residential demand for access to the Internet

**Paul N. Rappoport, Donald J. Kridel,
Lester D. Taylor, James H. Alleman and
Kevin T. Duffy-Deno**

INTRODUCTION

The focus in this chapter is on the residential demand for access to the
Internet, and represents an extension of earlier work on Internet access
demand by Rappoport et al. (1998), Kridel et al. (1999, 2000) and Duffy-
Deno (2000). The analysis of broadband demand has been studied by
Eisner and Waldon (1999), Madden et al. (1999) and Madden and
Simpson (1997). With the aggressive marketing of cable modems and
ADSL service, a growing number of residential households in the United
States (US) now have a choice regarding how they access the Internet. The
choice set available, however, is not uniform. In some areas the only form
of access is through dial-up modems, while in other areas various forms
of high-speed access (cable modems or ADSL) are also available. This
chapter reports the results from a set of models of Internet access where
the models are differentiated by the availability of Internet access options.
The models are based on the analysis of surveys submitted by over 20 000
households during the period January–March 2000.[1] Among other
things, broadband penetration rates are presented and compared to
Internet access estimates presented in the NTIA report (2000), 'Falling
through the net: Toward digital inclusion'.[2] In addition a more complete
set of elasticity estimates, for both basic and high-speed access to the
Internet, is provided. The chapter is organized as follows. A brief
summary of the demographics of Internet and broadband access is provided, enabling comparison of information obtained in the omnibus
survey with results presented in the NTIA (2000) report. The following
section presents the underlying models for estimation, then survey data
are discussed. The next section provides the results from the estimation.
Finally, results and implications are discussed.

DEMOGRAPHIC PROFILES OF ACCESS TO THE INTERNET

Consumer interest in the Internet is well documented. The NTIA (2000) study reports overall Internet penetration rates of 42.1 per cent in the US.[3] The survey data used in this study suggest household penetration rates of 46.5 per cent. These data represent a substantial increase in the proportion of households with access to the Internet. As recently as 1998, national studies indicated that only 26 per cent of households had access to the Internet.[4] Even in 2000, the vast majority of these households accessed the Internet through dial-up modems. However, the rate of broadband access is increasing. Recent estimates have 4.5 per cent of all households accessing the Internet through broadband connections. Or, over 10 per cent of those connected to the Internet access the Internet via broadband.[5] Estimates obtained from the omnibus survey indicate that 9.7 per cent of households that access the Internet have broadband or high-speed connections. This translates into 4.9 per cent of all households with broadband access. The reasons for the increasing popularity of broadband access are the availability and marketing of cable modems (CM) and ADSL services; the perceived need by users for more speed and bandwidth, and the decreasing price of broadband access. Figures 3.1 to 3.7 display Internet penetration rates according to household size, level of education, and income. As noted, these charts are based on data from an omnibus survey.[6] Where appropriate, NTIA results are presented for comparison.

Figure 3.1 demonstrates that Internet penetration increases with household size. The NTIA indicate that households with children have above average penetration, a finding consistent with Figure 3.1. Figure 3.2 shows broadband penetration by household size. The likelihood of broadband access increases with household size. Figure 3.3 demonstrates a positive correlation between education and Internet penetration. Having some college education is associated with above average penetration. Similar findings are found by the NTIA.[7] High-speed access and education in Figure 3.4 follow a similar pattern. Households with some college education have above average penetration rates. Previous studies show income positively correlated with Internet access and use.[8] Figure 3.5, which relates Internet penetration to income category, is consistent with this finding. Note the wealthiest households are almost four times more likely to have Internet access than poorest households. Figure 3.6 shows a similar pattern for broadband access. Income has a stronger effect on broadband than Internet access with the wealthiest households five and a half times more likely to have broadband access. Finally, the NTIA indicate a much higher broadband penetration at lower incomes than reported by the omnibus

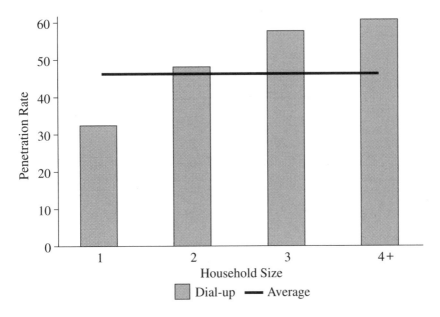

Figure 3.1   Internet penetration by household size

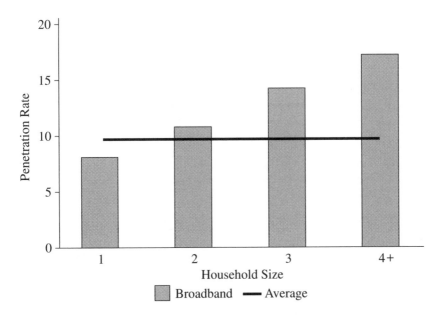

Figure 3.2   Broadband penetration by household size

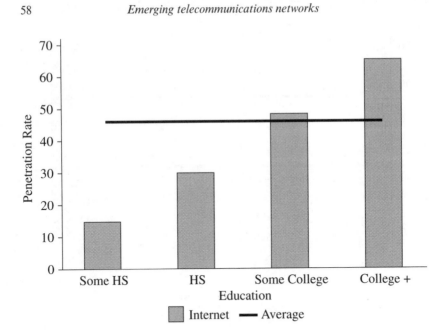

Figure 3.3   Internet penetration by education

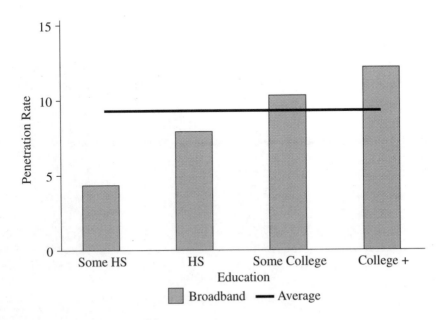

Figure 3.4   Broadband penetration by education

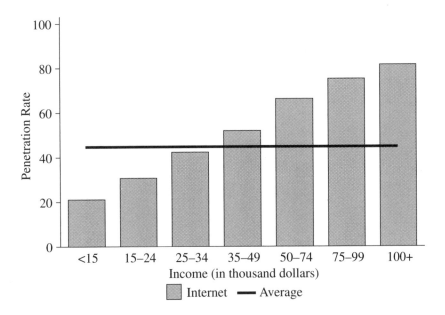

Figure 3.5 Internet penetration by income

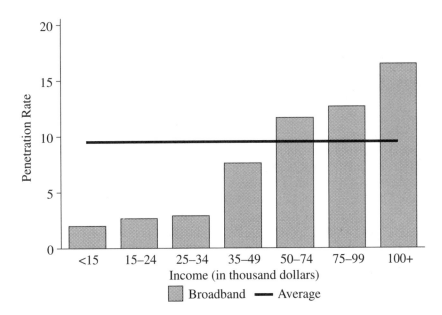

Figure 3.6 Broadband penetration by income

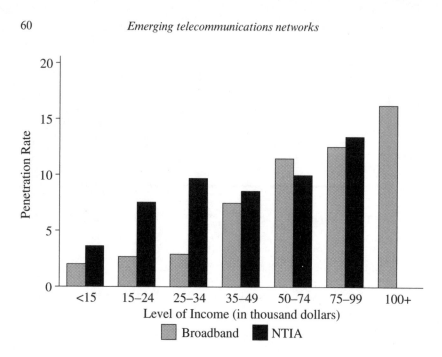

Figure 3.7   Broadband penetration by income, comparison with NTIA

survey. Figure 3.7 provides the comparison by income. The results should be interpreted with care since the sample sizes for broadband are small.

INTERNET ACCESS MODELS

Several models of residential Internet access are developed in the literature (Rappoport et al., 1998; Kridel et al., 1999, 2001). This study extends that work by allowing the choice of access to depend on the availability of the service. For broadband, this requires that either cable modem or ADSL service be available. Not all choices, however, are available to all households. Thus, for some regions, models are estimated to reflect the available full range of choice but, for other areas, the number and type of choice are limited. Table 3.1 defines areas by the available Internet options.

   Area 2 and Area 3 face different access options that imply different decision trees for estimation. For example, in Area 2 the possible decision (or tree) structures are displayed in Figure 3.8. For households in Area 1, the only choice option available is Internet access via dial-up access or no Internet access. A binary choice model is appropriate in this situation. In Area 2, the choice set is more complex as a broadband service is also avail-

Table 3.1   Area by available Internet access options

Area	Internet access options
Area 1	Dial-up only
Area 2	Dial-up and cable modem or ADSL
Area 3	Dial-up, cable modem and ADSL

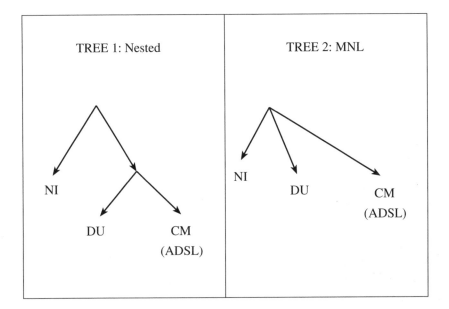

Note: NI is no Internet access. DU means dial-up access. CM is cable modem access. ADSL is ADSL access.

Figure 3.8   Area 2 decision structures

able. Households in Area 3 face even more options as both cable modem access and ADSL service are offered.

Models for Area 2 can be estimated with either a tree 1 or a tree 2 structure; that is, either by a nested multinomial logit (NMNL) (tree 1) or an unordered multinomial logit (MNL) specification (tree 2). The NMNL model assumes a choice between dial-up or some type of broadband access, given the household has Internet service. The MNL model is based on the mutually exclusive and exhaustive choices confronting the household; no Internet, dial-up or some type of broadband access. For Area 3, two

Emerging telecommunications networks

specifications for estimation are possible. First there is an NMNL structure. Here possible branches to consider are: Internet service provided by either dial-up or broadband service (for models in Area 3, both types of broadband service are available) and the type of broadband service chosen, assuming the household has broadband service. The first branch can also be specified as an MNL.

The MNL probability model where the only option for Internet access is dial-up access is described by equation (3.1):

$$\Pr(Y=1)=\frac{e^{\beta'x}}{\sum e^{\beta'x}} \tag{3.1}$$

where $Y=1$ is the choice of Internet service using dial-up access and $Y=0$ represents no Internet access. The choice of explanatory variables includes income, household size, education, age, gender and price. For tree 2 the available choices are dial-up and some high-speed alternative. For this situation the MNL model associated with tree 2 is utilized, and has the form:

$$\Pr(\text{choice}_j)=\frac{e^{\beta_j'x}}{\sum e^{\beta_j'x}}. \tag{3.2}$$

The choices considered are no Internet, dial-up Internet, and either cable modem or ADSL access. Two models are estimated. The first includes those observations where both dial-up access and cable modem service (but not ADSL service) are available. The second considers observations where both dial-up access and ADSL (but not cable modem service) are available.

Area 3 has all choices available. In this situation either the MNL (3.2) or the NMNL specification could be estimated. In the NMNL specification, the decision tree assumes individuals choose a branch (dial-up or broadband) and then an alternative, given the branch (for the branch broadband, the alternatives are ADSL or cable modem). The probability of choosing an alternative k from branch j and limb i is:[9]

$$\Pr(k|j,\,i)=\frac{e^{\beta'x_{k|j,i}}}{\sum_{n|j,i} e^{\beta'x_{n|j,i}}}=\frac{e^{\beta'x_{k|j,i}}}{e^{J_{j|i}}} \tag{3.3}$$

$$\Pr(j|i)=\frac{e^{\alpha'y_{j|i}+\tau_{j|i}J_{j|i}}}{\sum_{m|i} e^{\alpha'y_{m|i}+\tau_{m|i}J_{m|i}}} \tag{3.4}$$

$$\Pr(i)=\frac{e^{\gamma'z_i+\sigma_i}}{\sum e^{\gamma'z_i+\sigma_i}} \tag{3.5}$$

where $J_j|_i$ is the inclusive value for branch j on limb i, such that

$$Pr(k, j, i) = Pr(k|j, i)Pr(j|i)Pr(i). \qquad (3.6)$$

The NMNL model makes use of limbs for Internet access (access and no access), service branches (dial-up or broadband); and broadband alternative twigs (cable modem or ADSL). The difference between the NMNL using the explicit tree structure and the MNL specification is the specification of the tree structure. The NMNL assumes choices are based on perceived groups on the part of consumers.

DATA

The information used in the analysis is obtained from the Centris® Omnibus Survey administered by the Marketing Science Corporation of Fort Washington, PA. The survey covered the period January to March 2000. During that period, 20755 households were surveyed. Respondents were selected randomly. The Centris® survey contains questions on households' Internet use, the form of access and its cost. Information is also obtained on household computer use, as well as other electronic equipment, and household demographics. In addition to these survey data, additional information on cable modem availability and price, and ADSL availability and price are used. These data are from Duffy-Deno (2000). Price data are obtained partly from a national online survey of Internet households conducted in March 2000 by TNS Telecoms. The TNS survey targeted responses from each of three groups of Internet households across the US, that is, those that use dial-up, ADSL, or cable modem (CM) to access the Internet. Only households whose primary computer for Internet access is located in the home were solicited for the survey.

Online survey price data are augmented by household Internet access price data. For example, the price for ADSL and CM access, both inclusive of any ISP charge, that a given dial-up access household would be likely to have to pay for high-speed access (HSA), where the latter is available. ADSL prices, for those households who do not have ADSL access, is obtained by assuming prices as those of the household local exchange carrier (LEC), for example Pacific Bell.[10] CM prices, for those households who do not have CM access, are assumed to be those of the household CATV provider, for example Time Warner.[11] Finally, dial-up prices, for those households who do not have dial-up access, are assumed to be the average ISP price paid by survey households for the Census region in which the household is located.

These online survey data are also augmented by data on the availability of HSA in the surveyed areas. Clearly, a household may not have HSA because neither ADSL nor CM access is available. Using the zip code and telephone number provided by online survey respondents provides an indication of whether ADSL and/or CM access is being marketed in the household's immediate area. Five-digit zip codes are used to determine whether CM service is available within the household zip code. This was accomplished through checks of the CM-provider web sites. While cable modem service may not be available throughout a zip code, knowledge that CM service is available in at least some portion of a zip code suggests that cable operators are actively marketing the service in the immediate geographic region. Likewise, the household's area code and prefix were used to determine whether the LEC, for example Pacific Bell, is providing ADSL service from the central office servicing that area code and prefix. Again this is accomplished through an availability check on the LEC web sites, as well as through the use of data provided by some LECs as to which central offices are currently provisioned for ADSL. Distance limits and line requirements can preclude ADSL service to a household even when a serving central office is enabled to provide ADSL. However, as with cable modems, an indication can be obtained as to the geographic extent to which ADSL currently is being marketed. Consequently availability of HSA is defined in terms of whether ADSL and/or CM access is being marketed somewhere in a household's immediate geographic area, that is, the serving wire center or five-digit zip code.[12] Price and availability data are then matched to the Centris® household records at the zip-code level.

Variables and Delimiters

Descriptions of variables are given in Table 3.2.

EMPIRICAL RESULTS

Area 1

The Area 1 model is specified in terms of a logistic model of choice between having Internet access via dial-up modem ($Y=1$) or no Internet access ($Y=0$). For this model a sub-sample of 15387 respondents is identified of all records where either cable modem access or ADSL service was not available. The results from this estimation are displayed in Table 3.3.

As suggested by the demographic profiles, age, income, education level and household size are all highly significant. The categorical variables are

Table 3.2 Variable descriptions

Variable	Description
AGE	Age in years of the household head (years).
ADSLPRICE	Average price of ADSL service (US dollars).
BBPRICE	Average price of broadband service (dollars).
CPRICE	Average price of cable modem service (dollars).
DUPRICE	Average price for dial-up Internet service (dollars).
CMAV	Cable modem availability; = 1 if available; zero otherwise.
ADSLAV	ADSL availability; = 1 if available; zero otherwise.
EDUC (i)	Highest educational qualifications attained by the household head. Measured as a categorical variable for the groups (i): no high school, high school, some college or college graduate.
GENDER (j)	Gender of household head; j = 1 if male; zero otherwise.
HHSIZE (k)	Number of household residents; k = 1, 2, ..., 7.
INCOME (l)	Household income is measured as a categorical variable with bands in thousand dollars: l = <15, 15–24, 25–34, 35–49, 50–74 and 75–99).

Table 3.3 Estimation results: no Internet or dial-up access model

Variable	Coefficient	Std error	t-statistic	p value
Constant	2.891	0.515	5.60	
DUPRICE	−0.036	0.020	−1.79	0.035
HHSIZE (1)	−0.370	0.198	−1.86	0.630
HHSIZE (2)	0.221	0.195	1.11	0.911
HHSIZE (3)	0.279	0.196	1.42	0.155
HHSIZE (4)	0.378	0.197	1.92	0.055
HHSIZE (5)	0.394	0.201	1.96	0.495
INCOME (<15)	−1.699	0.093	−18.26	0.000
INCOME (15–24)	−1.395	0.082	−17.01	0.000
INCOME (25–34)	−1.019	0.078	−13.06	0.000
INCOME (35–49)	−0.729	0.073	−9.99	0.000
INCOME (50–74)	−0.325	0.074	−4.39	0.000
INCOME (75–99)	−0.100	0.084	−1.19	0.233
EDUC (<HS)	−1.572	0.100	−12.57	0.000
EDUC (HS)	−0.938	0.071	−13.15	0.000
EDUC (Some College)	0.379	0.072	5.26	0.000
EDUC (College)	0.530	0.072	7.36	0.000
GENDER (M)	0.831	0.358	2.32	0.020
AGE	1.955	0.127	15.39	0.000

all relative to the value for their excluded categories, that is, EDUC (post college), HHSIZE (over 6), INCOME (over 100000). The DUPRICE access elasticity is −0.372.[13] This compares quite closely with −0.38 elasticity reported in Kridel et al. (1999).

Area 2

In this model the sub-sample includes 3705 observations for which cable service is and ADSL service is not available. The average price for cable service is USD40.57, while the average price for dial-up service in this sample is USD20.11. The output from this estimation is displayed in Table 3.4. Estimated own and cross-price elasticity are reported in Table 3.5.[14] It is useful to note that for the choice of dial-up, cable modem service is a significant substitute. However, for those customers with cable modem service,

Table 3.4   Estimation results: dial-up or cable modem model

Variable	Coefficient	Std error	t-statistic	p value
$p(Y=\text{dial-up})$				
Constant	0.643	0.904	0.711	
DUPRICE	−0.023	0.010	−2.310	0.027
CPRICE	0.259	0.018	14.714	0.000
INCOME (<15)	−0.049	0.010	−0.491	0.623
INCOME (15–24)	−0.079	0.078	−1.001	0.340
INCOME (25–34)	−0.030	0.042	−0.710	0.453
INCOME (35–49)	0.043	0.025	1.720	0.068
INCOME (50–74)	0.057	0.026	2.192	0.051
INCOME (75–99)	0.074	0.027	2.740	0.005
GENDER (M)	0.328	0.177	1.850	0.064
AGE	−0.028	0.008	−3.014	0.003
$p(Y=\text{cable modem})$				
Constant	−1.894	1.780	−1.059	
DUPRICE	0.004	0.029	0.013	0.986
CPRICE	−0.024	0.009	−2.622	0.008
INCOME (<15)	−1.358	0.507	−2.677	0.007
INCOME (15–24)	−1.151	0.396	−2.905	0.004
INCOME (25–34)	−1.227	0.352	−3.486	0.000
INCOME (35–49)	−1.125	0.290	−3.882	0.000
INCOME (50–74)	−0.641	0.243	−2.640	0.008
INCOME (75–99)	−0.736	0.286	−2.574	0.010
GENDER (M)	0.652	0.177	3.684	0.000
AGE	0.049	0.021	2.333	0.024

Table 3.5   Area 2 price elasticity estimates

	Dial-up	Cable modem
Dial-up	−0.230	0.518
Cable modem	0.001	−0.895

dial-up service is insignificant. Also there appears to be a more substantial impact of income in the choice of service for cable modem. There are 4041 observations available for the model with choices of dial-up access or ADSL service, but not cable modem access. The output from this estimation is displayed in Table 3.6. Estimated and cross-price elasticities are reported in Table 3.7.

Area 3

In this area, consumers face the complete choice set of no access, dial-up, cable modem and ADSL services. A sub-sample of 5255 observations is used in the estimation of this model. An MNL model is specified to estimate the choice between dial-up access and broadband service. The branch is estimated as an NMNL model utilizing the LIMDEP NLOGIT sub-routine.[15] The correct specification of the branches depends on the significance of the inclusive values that link branch to twig. NLOGIT provides for the complete specification and estimation of the inclusive values.[16] The following Table 3.8 reports the estimation of the branch in the structure, that is, the choice between dial-up access and broadband access.

The MNL model confirms that INCOME and EDUC are major determinants of choice. The increased importance of income, college and post-college education in the choice of broadband service is also evident. The inclusive value as computed from the lower branch is significant at the 5 per cent level, confirming the tree structure is adequate. The derived elasticities at this stage are given in Table 3.9.

Focusing on the twig, unobservable differences in the utility functions for CM and ADSL are unlikely to be related to any of the observed demographic variables. Hence only choice-specific variables are included (prices) to model broadband choice. The following output in Tables 3.10 and 3.11 pertains to the estimation of the choice between cable modem and ADSL access; PRICEDIFF is the difference in the prices facing the subscriber.

Table 3.6 Estimation results: dial-up or ADSL model

Variable	Coefficient	Std error	t-statistic	p value
$p(Y=\text{dial-up})$				
Constant	1.575	0.161	9.749	
DUPRICE	−0.016	0.002	−7.490	0.000
ADSLPRICE	0.961	0.250	3.844	0.000
INCOME (<15)	−0.409	0.046	−8.545	0.000
INCOME (15–24)	−0.288	0.042	−6.927	0.000
INCOME (25–34)	−0.219	0.036	−6.027	0.000
INCOME (35–49)	−0.111	0.033	−3.444	0.000
INCOME (50–74)	−0.051	0.031	−1.761	0.078
INCOME (75–99)	0.031	0.032	0.912	0.362
GENDER (M)	0.031	0.018	1.701	0.089
AGE	−0.004	0.001	−5.312	0.000
EDUC (<HS)	−0.219	0.055	−3.987	0.001
EDUC (HS)	−0.078	0.035	−2.223	0.026
EDUC (S College)	0.013	0.033	0.390	0.697
EDUC (College)	0.046	0.021	2.190	0.032
$p(Y=\text{ADSL})$				
Constant	6.927	1.117	6.204	
DUPRICE	0.005	0.012	0.368	0.712
ADSLPRICE	−0.031	0.007	−4.428	0.000
INCOME (<15)	−2.402	0.346	−6.939	0.000
INCOME (15–24)	−1.898	0.321	−5.906	0.000
INCOME (25–34)	−1.582	0.299	−5.286	0.000
INCOME (35–49)	−0.996	0.288	−3.464	0.000
INCOME (50–74)	−0.588	0.293	−2.002	0.043
INCOME (75–99)	0.571	0.399	1.432	0.152
GENDER (M)	0.212	0.132	1.541	0.000
AGE	−0.021	0.005	−4.602	0.000
EDUC (<HS)	−1.067	0.361	−2.892	0.038
EDUC (HS)	−0.433	0.264	−1.637	0.101
EDUC (S College)	0.129	0.272	0.488	0.625
EDUC (College)	0.502	0.272	1.841	0.066

Table 3.7 Area 2 price elasticity estimates

	Dial-up	ADSL
Dial-up	−0.168	0.423
ADSL	0.040	−1.364

Table 3.8   Estimation results: dial-up or broadband model

Variable	Coefficient	Std error	t-statistic	p value
$p(Y = \text{dial-up})$				
Constant	0.474	0.448	1.06	
DUPRICE	−0.285	0.021	−1.34	0.183
BBPRICE	0.017	0.008	2.13	0.024
INCOME (<15)	−0.854	0.071	−12.03	0.000
INCOME (15–24)	−0.557	0.058	−9.67	0.000
INCOME (25–34)	−0.191	0.052	−3.67	0.000
INCOME (35–49)	0.066	0.047	1.41	0.159
INCOME (50–74)	0.316	0.049	6.45	0.000
INCOME (75–99)	0.507	0.048	10.40	0.000
GENDER (M)	0.246	0.031	7.85	0.000
AGE	−0.012	0.001	−12.00	0.000
EDUC (<HS)	−1.687	0.083	−20.44	0.000
EDUC (HS)	−0.955	0.057	−16.86	0.000
EDUC (S College)	−0.311	0.058	−5.30	0.000
EDUC (College)	0.137	0.059	2.31	0.021
$p(Y = \text{broadband})$				
Constant	6.927	1.117	6.204	
DUPRICE	0.002	0.061	0.037	0.982
BBPRICE	−0.039	0.010	−3.90	0.000
INCOME (<15)	−1.202	0.284	−4.243	0.000
INCOME (15–24)	−0.972	0.220	−4.411	0.000
INCOME (25–34)	−0.715	0.189	−3.787	0.000
INCOME (35–49)	−0.535	0.158	−3.375	0.000
INCOME (50–74)	0.244	0.132	1.856	0.064
INCOME (75–99)	0.751	0.152	4.917	0.000
GENDER (M)	0.896	0.097	9.177	0.000
AGE	−0.018	0.002	−9.000	0.000
EDUC (<HS)	−2.291	0.316	−7.229	0.000
EDUC (HS)	−1.437	0.159	−9.010	0.000
EDUC (S College)	0.316	0.142	2.222	0.056
EDUC (College)	0.499	0.142	3.511	0.000
Inclusive value	0.445	0.227	1.960	0.025

Table 3.9   Area 3 price elasticity estimates

	Dial-up	Broadband
Dial-up	−0.277	0.725
Broadband	0.021	−1.491

Table 3.10   Estimation results: cable modem or ADSL

Variable	Coefficient	Std error	t-statistic	*p* value
PRICEDIFF	−0.041	0.009	−4.49	0.000

Table 3.11   Area 3 price elasticity estimates

	Cable modem	ADSL
Cable modem	−0.587	0.766
ADSL	0.618	−1.462

CONCLUSIONS

Using recent data from an omnibus survey, models of consumer access to the Internet are estimated. The models cover situations where choice is dial-up access only, dial-up or broadband access and broadband but cable modem or ADSL. The dial-up price elasticity consistently falls in the range [−0.4 to −0.2], which accords with available estimates. The results confirm that availability of choice matters. When the only alternative is ADSL, a significant cross-price effect suggesting ADSL is a strong substitute for dial-up access is noted. This ADSL price elasticity is greater than 1 in magnitude. However, as the penetration rate of ADSL service increases, the price elasticity becomes inelastic. A similar result is found in the model where cable modems are the only alternative to dial-up access. The cable modem price elasticity is smaller in absolute value than the ADSL price elasticity, reflecting the greater penetration of cable modems in the residential market. For the model where all types of access are available to the household, both ADSL and cable model service are strong substitutes for dial-up access. Also, not surprisingly, ADSL and cable modems are substitutes. There are other factors at work in the decision to adopt broadband service. These are summarized in the type, duration and reach of household usage. The logical extension of the choice modeling requires the collection and use of Internet activity information. This information is being collected and will be incorporated into future broadband models. These broadband penetration data suggest that the market for broadband services will continue to grow. As the demand for bandwidth and service quality increases, the market should witness the migration of customers from dial-up to broadband services. This migration will accelerate as broadband prices, relative to dial-up prices, fall, the availability of broadband service increases and applications that use broadband services expand.

NOTES

The authors wish to thank Dr Don Michels and Dr Dale Kulp for their assistance.

1. Centris® Omnibus Survey, Marketing Science Corporation, January–March 2000.
2. NTIA, 'Falling through the net: Toward digital inclusion', US Department of Commerce, NTIA, 2000 at http://www.ntia.doc.gov/ntiahome/digitaldivide/.
3. *Op. cit.*, NTIA, pp. 1–3.
4. Ibid., NTIA, p. 1.
5. Ibid., pp. 23.
6. Centris® Market Systems Group, Fort Washington, PA.
7. *Op. cit.*, p. 99.
8. See Rappoport et al. (1998) and Kridel et al. (1999) and (2001).
9. See Greene (2000: 865–72).
10. The price of the serving LEC's entry residential ADSL offering is used. The LECs were the primary source of ADSL service to the residential sector during the time period covered by the survey. These prices were obtained from the LECs' web sites.
11. The CM access prices are those for current CATV subscribers, obtained from the CATV provider web sites.
12. To determine whether ADSL or CM access is available to a specific household premise would require the input of the household's address into the service providers' availability checks, offered on their web sites. The online survey did not ask for a premise address.
13. The formula used is $\beta \cdot \text{DUPRICE}(1 - \text{share}_{\text{Internet}})$ where the mean DUPRICE is USD20.38.
14. Own-price elasticity is calculated by the rule $\beta_j \cdot \text{DUPRICE}(1 - \text{share}_j)\text{PRICE}_j$. and the cross-price elasticity is calculated by applying the rule $\beta_i \cdot \text{DUPRICE}(1 - \text{share}_i)\text{PRICE}_j$.
15. Greene, LIMDEP Version 7.0.
16. The last branch, ADSL versus cable modem is specified as a function of price only, there being no other choice-specific attributes available. The second branch in tree 3 is viewed in terms of an NMNL.

REFERENCES

Duffy-Deno, K.T. (2000), 'Demand for high-speed access to the Internet among Internet households', TNS Telecoms, 2000.

Eisner, J. and Waldon, T. (1999), 'The demand for bandwidth: Second telephone lines and on-line services', Federal Communications Commission, Washington, DC, forthcoming in *Information Economics and Policy*.

Greene, W.H. (2000), *Econometric Analysis*, 4th edn, Upper Saddle River, NJ, Prentice Hall.

Kridel, D.J., Rappoport, P.N. and Taylor, L.D. (1999), 'An econometric model of the demand for access to the Internet', in Loomis, D.G. and Taylor, L.D. (eds) *The Future of the Telecommunications Industry: Forecasting and Demand Analysis*, Boston, Kluwer Academic Publishers.

Kridel, D.J., Rappoport, P.N. and Taylor, L.D. (2001), 'An econometric model of the demand for access to the Internet by cable modem', in Loomis, D.G. and Taylor, L.D. (eds) *Forecasting the Internet: Understanding the Explosive Growth of Data Communications*, Boston, Kluwer Academic Publishers.

Madden, G., Savage, S.J. and Coble-Neal, G. (1999), 'Subscriber churn in the Australian ISP market', *Information Economics and Policy*, 11(2), 195–208.

Madden, G. and Simpson, M. (1997), 'Residential broadband subscription demand: An econometric analysis of Australian choice experiment data', *Applied Economics*, 29, 1073–8.

NTIA (2000), 'Falling through the net: Toward digital inclusion', US Department of Commerce, National Telecommunications and Information Administration, Washington DC, October.

Rappoport, P.N., Taylor, L.D., Kridel, D.J. and Serad, W. (1998), 'The demand for Internet and on-line access', in Bohlin, E. and Levin, S.L. (eds) *Telecommunications Transformation: Technology, Strategy and Policy*, Amsterdam, IOS Press, 205–18.

4. Electronic commerce and industrial organization

Steven Globerman

INTRODUCTION

The emergence and growth of electronic commerce (e-commerce) are seen as momentous developments by many observers, on a par with the industrial revolution in terms of the potential economic consequences. The prospective impact of e-commerce on the industrial organization of product markets has in particular been noted. Specifically, e-commerce is seen as encouraging profound changes in the geographical markets of many products, as well as associated changes in the structural competitiveness of many product markets. This study identifies and evaluates the major potential impacts of e-commerce on industrial organization. In particular, it addresses whether and to what extent geographical distance, and the related costs of transacting over geographical distance, will become substantially less important over time for different sets of producers and consumers. It also considers whether traditional barriers to entry are becoming less important in conditioning the degree and nature of competition in relevant product markets. Further, the potential for new forms of market intermediation to emerge, and traditional forms to disappear, is addressed.

The following section provides a brief description of e-commerce, its current status and prospects for the future. Identification and assessment of the main hypotheses linking the growth of e-commerce to changes in organizational characteristics of affected industries are then made. Evidence bearing on the main hypotheses is also evaluated. A discussion of policy implications concludes the study.

DESCRIPTION OF E-COMMERCE

Definition of E-commerce

The term e-commerce is used to describe many uses of modern telecommunications and information technology. The most encompassing definition

73

of e-commerce includes any form of business activity conducted on the electronic medium (Wigand, 1997). This would include electronic data interchange (EDI), electronic mail and related types of communication. While EDI has been explicitly equated with e-commerce in the past, it is more properly viewed as a subset of e-commerce. In the vernacular, EDI encompasses business-to-business (B-to-B) electronic transactions. Commercial transactions involving sales to households are identified as business-to-consumer (B-to-C) electronic transactions. Non-commercial transactions conducted electronically do not qualify as e-commerce by most definitions, although, as a practical matter, the margin between commercial and non-commercial transacting is somewhat vague. Also, commercial transactions must be carried out over the Internet/World Wide Web (WWW) to be typically characterized as e-commerce. While many commercial transactions are carried out on private electronic networks, the main hypotheses linking e-commerce to industrial organizational changes focus on public access networks, of which the WWW is the dominant model.

The Magnitude and Nature of E-commerce

Systematically collected data on the magnitude and nature of e-commerce transactions are generally unavailable. Instead, one must use *ad hoc* observations and estimates about the use of the WWW for commercial purposes, and those observations and estimates apply primarily to the United States (US). The embryonic nature of B-to-C e-commerce is underscored by the data summarized in Table 4.1. The table reports estimates of purchases on the WWW by US residential customers in different product categories for 1997 along with projections for the year 2000. The projection for total Internet shopping in the year 2000 already appears to be a serious underestimate, as US B-to-C sales in 1998 are estimated to have been almost US$ (USD) 8 billion, and USD31 billion in 1999 (Baker and Echikson, 2000).[1] Forecasts of future of B-to-C e-commerce volumes project rapid and even accelerating growth. For example, Forrester Research Inc. projects that B-to-C commerce will grow to USD108 billion in 2003.[2]

It should be noted that most forecasts of electronic commercial activity do not restrict their definition to transactions that involve on-line payments. In other words, they include transactions in which the consumer merely identified and/or ordered a product on-line, while paying for the purchases off-line.[3] Even without this qualification, the optimistic projections of household purchases on the WWW still suggest that conventional purchasing channels will remain dominant. This is apparent from 1999 total personal consumption expenditure in the US of nearly USD6.3 trillion. B-to-C electronic commerce in other parts of the world has grown

Table 4.1   Estimates of Internet shopping (million USD)

	1997	2000
PC hardware and software	863	2901
Travel	654	4741
Entertainment	298	1921
Books and music	156	761
Gifts, flowers and greeting	149	591
Apparel and fashion	92	361
Food and beverage	90	354
Jewelry	38	107
Sporting goods	20	63
Consumer electronics	19	93
Other	65	197
Total	**2444**	**12090**

Source: 'The virtual mall gets real', *Business Week*, 26 January 1998, pp. 90–91.

more slowly than in the US. For example, Europeans bought only USD5.4 billion worth of goods on-line according to one consultant's estimate (Baker and Echikson, 2000).

The product scope of early B-to-C e-commerce has been narrow, as is suggested by Table 4.1. In particular, on-line computer hardware and software sales constitute a disproportionately large share of household products purchased using the WWW. On-line purchasing in the areas of travel, entertainment, and books and music has also been prominent. Recently there has been some evidence of a broadening of product categories for which household use of the WWW is growing. For example, there has been an increase in sales of previously laggard categories such as clothing, furniture and groceries. This development is apparently due, in part, to women doing more on-line shopping.[4] Nevertheless, computers and software, books and music, and travel remain the dominant product categories for Internet retail activity.[5]

Rapid growth in B-to-B commerce is also anticipated, notwithstanding its earlier introduction compared to other e-commerce applications. In fact, the historical adoption of B-to-B e-commerce has been relatively slow. For example, it is estimated that around 3000 US firms were using EDI in 1986,[6] and the number had increased to only around 100000 in 1994. Adoption of EDI in Europe was even slower. It is estimated that in 1988, the number of European (including UK) firms using EDI was around 2400, and the estimated number increased to around 20000 in 1994 (Pfeiffer, 1995).

Nevertheless, B-to-B e-commerce appears to be more substantial than B-to-C commerce. For example, it is estimated that US business exchanged around USD17 billion in goods and services in 1998 through e-commerce.[7] An important recent development in B-to-B e-commerce is the use of group WWW sites for buying and selling products. An example is the recently announced decision by six major US airline carriers to establish a Web site to sell discounted tickets on-line.[8] Another is a plan announced by the three biggest PC makers to operate an electronic exchange to purchase components and to sell surplus parts (McWilliams, 2000). The implications of multi-firm Web sites for industrial organization are considered below.

HYPOTHESIZED IMPACTS

Hypothesized impacts of e-commerce on the organizational characteristics of industry include the expansion of geographical markets and a corresponding decrease in structural concentration. Relevant geographical markets may be thought of as the smallest geographical area within which any seller (or group of sellers) can impose and sustain a profitable price increase of at least 5 per cent. As a practical matter, the boundaries of a relevant geographical market are set by a variety of transportation and related transactions costs that make it unprofitable for producers located beyond the boundaries to ship into the market, even after a 5 per cent price increase in the market. Equivalently, transportation and transactions costs make it uneconomical for consumers to search for alternative supply outside the region. Costs of transacting are essentially composed of costs of physically searching for products, sellers and buyers (search costs); costs of establishing and fulfilling contracts (contracting costs); costs of ensuring that the terms of contracts are met (monitoring costs); and costs incurred in making changes to contracts over time (adaptation costs).[9]

Search Costs

Most discussions on the economic advantages of e-commerce have focused on the reduction of search costs resulting from the increased ease with which information about prices, product availability and demand can be obtained using the Internet. Such reductions in search costs are suggested to be especially relevant for specialized products for which market participants are few in number and possibly widely separated in geographical space.[10] As a consequence, markets for products whose search costs are significantly reduced by e-commerce should become more competitive, since a greater number of participants (hitherto separated by geographical space) will compete for favorable terms and conditions. As a further consequence, prices of products should more closely approximate their marginal costs, thereby contributing to improved allocative efficiency.[11]

Economists identify a search good as one where product attributes can be identified by inspection prior to purchase. Within this broad definition, search includes visual and tactile inspection. Computer equipment is an obvious example of a search good, as technical specifications are quite meaningful and are easily communicated to potential buyers. Financial securities listed on major stock exchanges are also search goods in that properties such as price, volume, dividend yield and so forth can be readily and easily determined prior to purchase. If the consumer must utilize a product in order to determine key attributes (such as ease of use and durability), the product is said to have 'experience' qualities.[12] An example is the automobile – subjective characteristics of an automobile, such as 'road feel' and steering responsiveness, can usually best be established by actually driving the car.

Critical attributes of some specialized products may not be confidently identifiable even after utilization. Economists identify such products as 'credence' goods, for example medical services. Since patients frequently improve their state of well being independently of the services of health care professionals, only fairly prolonged experience with the ministrations of one's health care specialist will provide the patient with insight into whether or not the specialist adds significant value to the patient's efforts to be healthy. Moreover, health problems and concerns are, to some extent, idiosyncratic. Hence, one patient's satisfactory experience with a health care professional may not be an absolutely reliable signal to other patients of the latter's likely satisfaction with that professional.

It is widely acknowledged that the Internet is a useful tool for collecting information about search goods. Indeed, the fact that purchasing computer equipment and travel and brokerage services has been prominent in the early e-commerce experience attests to the advantages enjoyed by those selling search-type goods on the WWW. Since price is an important searchable feature, the emergence of price-searching software will further enhance the advantages of electronic buying and selling of search goods. The ability to electronically download free samples of certain types of experience goods expands the scope for e-commerce to encompass many of these types of goods, as well. For example, music and book publications, software, financial information and advice and educational courseware, among other things, can be downloaded to enable potential buyers to evaluate products on offer. Increasingly, software on the WWW is capable of portraying product features in a context that more closely approximates personal inspection. For example, three-dimensional software images allow potential purchasers of real estate to take on-line tours of the interiors of houses. Similarly, buyers of designer clothing can have the fitting done electronically, using scanners, and can be provided with pictures of how they will look wearing the designed product.

For experience goods that cannot be effectively sampled electronically, producers can try to reassure consumers about qualitative attributes of the products in more traditional ways, such as by investing in the creation of brand names and by offering product satisfaction warranties. In this regard, it is unclear how the Internet, *per se*, will affect the costs that producers need to incur in order to create the trust capital required to make their quality claims credible to potential customers. Traditionally, large accumulated sunk costs in brand names and trademarks have been a major hostage that firms have made available to potential consumers in order to create trust (Klein and Leffler, 1981). To the extent that the Internet allows firms to lower (or avoid) the sunk cost investments that have been traditionally required to market and promote experience goods, consumers may become even more concerned about purchasing 'lemons' through the Internet, and the marketing of experience goods may not be significantly affected by the emergence of e-commerce.[13] It is technically possible for credence goods to be distributed over the WWW, for example psychologists are selling their services on-line to clients primarily through electronic mail. Medical doctors can also be contacted at Web sites to answer health care-related questions, although diagnoses are usually qualified to minimize the risk of litigation.[14] Nevertheless, the individual's purchase of credence goods is likely to remain strongly guided by recommendations from close contacts such as family, friends and other professionals.[15]

To sum, it is likely to be traditional search goods whose transaction costs are most substantially reduced by e-commerce. Table 4.1 generally confirms that search goods and relatively inexpensive experience goods that can be electronically sampled comprise the majority of products that have been purchased on-line. As consumers become more confident about both the security of Internet payment systems and the willingness and ability of on-line merchants and auctioneers to ensure the quality of the products being transacted through their WWW sites, more expensive search goods such as real estate, collectibles and luxury consumer items, should become more frequently purchased using the WWW.

Other Transaction Costs

There has been less discussion, and much less of a consensus formed, regarding the impact of e-commerce on other types of transaction costs. It has been argued that the widespread adoption of standardized electronic contracts will lower the average costs of contracting. This is especially so for B-to-B transactions, since a repetitive activity with relatively high variable costs would be replaced by a once-and-for-all effort with relatively high fixed (and sunk) costs but low variable costs. However, besides the unresolved issues currently surrounding electronic contracts, it is unclear that transactions between parties, including those who regularly do business together, are sufficiently standardized that the need for contract modifications on an ongoing, and perhaps unpredictable basis, will be obviated. This caveat is especially relevant for international transactions where differences in legal regimes, contractual customs and so forth may oblige firms to enter into multiple agreements with a resulting loss of opportunities to standardize.

To the extent that perceived risks of opportunistic behavior are no lower for electronic commercial activities than for conventional commercial activities, electronic contracts may need to be complex and adapted frequently over time. E-commerce may, therefore, do little to reduce the costs associated with maintaining and enforcing contracts, including litigation costs. On the other hand, to the extent that the growth of e-commerce significantly expands the relevant geographical markets for many products, buyers and sellers of those products should experience lower costs associated with switching transaction partners. Lower switching costs, in turn, should reduce incentives for individual market participants to act opportunistically, all other things constant, which, in turn, should reduce the costs of establishing, maintaining and enforcing contracts.[16]

To sum, the focus of current discussions linking e-commerce to global competition has been on anticipated reductions in transaction costs related to dealing with multiple market participants in geographical space. The dominant hypothesis has been that e-commerce will lead to a dramatic reduction in search costs that, in turn, will substantially expand the geographical scope over which transactions can efficiently take place.[17] However, search costs are only one component of transaction costs, and it is unclear whether e-commerce will contribute to economizing on other transaction costs components. Moreover, reductions in search costs may be relatively unimportant for a range of products that must be inspected, or even utilized, to provide sufficient information to prospective buyers. Consequently, substantial improvements in technology and/or institutional arrangements may be required to ensure that a wide range of goods and services enjoy substantial reductions in the costs of transacting electronically.

Competition and Contestability

Industrial organization economists view competition as both a structural and a behavioral phenomenon. Structurally competitive markets are characterized by low ownership concentration. That is, the largest firms in the market enjoy relatively small market shares. Moreover, there are numerous competitors. Behaviorally competitive markets are characterized by vigorous price and non-price competition with rivals largely abstaining from what might be considered cooperative behavior.[18] Contestability is concerned with the influence that potential entry has on the behavior of existing competitors. In a contestable market, the threat of entry is sufficiently compelling that incumbent sellers are obliged to behave in a competitive manner, regardless of existing levels of ownership concentration. Indeed, in a perfectly contestable market, the equilibrium price and output will correspond to those that would obtain under perfect structural competition, even if only one seller exists in the market.[19] Relatively low (sunk) costs of entry and exit are the critical features of a contestable market.

The preponderant view of e-commerce is that it will promote increased competition. As noted earlier, the anticipated expansion of relevant geographic markets should reduce structural concentration by expanding the geographical scope over which firms can economically compete. E-commerce has also been suggested to reduce barriers to entry, especially for small firms, thereby enhancing contestability. In particular, e-commerce is allegedly characterized by much lower required sunk cost investments than is entry associated with more traditional distribution channels. As a case in point, Solomon (1995) suggests that it costs as little as USD1000 annually to open an electronic storefront on the Internet that is accessible by as many as 20 million persons, however, especially with the proliferation of WWW sites, it is becoming increasingly difficult to gain 'visibility' on the WWW. In order to gain easier access to individuals browsing the Internet, many on-line merchants have begun using the heavy traffic net search engine sites (the most frequently visited sites) as a springboard to their sites. Moreover, on-line shopping tools that can scour the entire WWW, and the many merchants selling goods on the WWW, are still fairly primitive. The growing premium on the visibility of a seller's Web site is arguably enhancing the competitive advantages of sellers that use a 'clicks-and-bricks' approach to Internet marketing. The latter refers generally to the leveraging of brand names created in conventional merchandising channels to promote selling efforts through the WWW. For example, the main on-line sports Web sites are attempting to increase viewers by cross-promoting with 'main events' they feature on other media, such as sister cable channels. As another example, an online auctioneer (Free Markets) of molded plastic parts and fasteners is using its physical presence in industrial Pittsburg as a basis for making its on-line auction site credible to market participants (Aeppel, 2000).

Brand name spillovers from conventional distribution and media channels can be expected to increase the sunk costs of entry for *de novo* sellers by obliging the latter to invest substantial amounts of money in creating a unique brand name. For example, several large Canadian banks apparently feel obliged to enter the US market through a combination of physical branches and Internet operations (Greenberg, 1999). This obligation is probably more relevant for sellers engaged in B-to-C e-commerce than for those engaged in B-to-B commerce. Nevertheless, even in the latter case, a reputation for being a reliable supplier is ordinarily required to gain access to the purchasing networks which are increasingly being formed by leading companies across a wide range of industries. In some cases, acceptance into the purchasing networks requires would be suppliers to have an established reputation for reliability or a demonstrated capacity to meet supply commitments. To the extent that such requirements impose delays on smaller entrants achieving minimum efficient size and/or oblige new firms to enter on a relatively large scale, with commensurately large sunk costs, e-commerce might not be the boon to contestability that some enthusiasts have contended. In other cases, incumbent firms may be able to use e-commerce capabilities to augment first-mover advantages. An example is American Airlines program to offer frequent fliers one-to-one marketing software. With this software, preferred customers can streamline their booking process by creating a profile of their home airport, seating and meal preferences and so forth.[20] Another example is the effort of established brokers such as Merrill Lynch to bundle personalized advisory services with on-line trading as an integrated service offering in response to the emergence and growth of discount on-line trading services.

The view that reductions in search costs will lead to more competitive pricing might also be too sanguine. For example, Picot et al. (1997) contend that electronic mediation of market transactions will not automatically lead to reduced prices compared to conventional market organizations, since sellers can be expected to implement various strategies to reduce market transparency in order to preserve previous price and income levels. For example, they may quote prices on the WWW as the basis for further negotiation, rather than as firm offers that will be filled if buyers meet the quoted price. In this way, some price discrimination remains possible based on the buyer's urgency for the product, his or her opportunity costs of the time spent haggling over the product and so forth. The use of bundled pricing and complicated charging schedules can also obscure price differences among sellers.[21]

The growth of industry group Web sites, as described above, might facilitate non-competitive pricing by dominant sellers or buyers who comprise the group. To the extent that industry group Web sites enjoy a reputation for guaranteeing product quality and distribution reliability, areas of bottleneck monopoly or monopsony power may emerge on the WWW. To be sure, in certain areas of B-to-B and B-to-C commerce, established multiproduct 'e-tailors', such as Amazon.com, will exist as competitors to some industry group Web sites; however, this source of competition will be less robust, the more technically specialized the set of products being sold. Likewise, auction sites, such as Ebay, may not be seen as reliable alternatives to industry-run sites for many would-be buyers or sellers of products for which small deviations from desirable specifications would render those products undesirable.

To sum, while surface considerations suggest that e-commerce will strongly enhance the competitiveness and contestability of markets, with associated gains in consumer welfare, there are grounds for concern that the early enthusiasm surrounding this prospective development was excessive, if not unjustified. At the least, pockets of market dominance may be cultivated on the WWW through combinations of purchasing and selling groups. As well, firms that enjoy positions of market dominance in conventional marketing channels may be able to use e-commerce initiatives to further entrench, or better exploit, their market dominance.

E-commerce and Disintermediation

The early popular view that e-commerce will lead to widespread disintermediation as a result of producers selling directly to customers is also coming in for some rethinking. This development reflects a growing recognition that the demand for information is not necessarily reduced by e-commerce. Indeed, it may well be increased by the growing opportunities for buying and selling on the WWW. In this context, depending on the activity in question, the demand for specialized expertise to facilitate transactions might either increase or decrease. For example, knowledge about how to execute transactions where the parties know and trust each other is being rapidly devalued by e-commerce. A case in point is stock brokerage. Simple executions of buy-and-sell orders are increasingly being performed on-line without the intermediation of traditional brokerage personnel.[22] At the same time, a growing number of companies are emerging to advise consumers on the relative merits of alternative electronic brokers. In short, there is a rapidly growing demand for intermediaries who can enable sellers to better identify and serve buyers on the WWW, as well as those who can assist buyers to locate suppliers in possession of products with specific characteristics and selling at competitive prices. Established intermediaries that have gained their expertise in conventional marketing media may or may not be successful in extending their acknowledged expertise to e-commerce activities. For example, traditional market research companies may have no sustainable competitive advantage in acquiring and processing sales and marketing data related to e-commerce for the benefit of competing as consultants to on-line sellers. On the other hand, where successful intermediation primarily draws on idiosyncratic and largely uncodifiable knowledge of specific markets, incumbent intermediaries could enjoy strong competitive advantages in supplying intermediation services on the WWW. For example, while it is relatively easy to imagine Netscape or Microsoft opening up their own electronic brokerage services, presuming they could obtain government approval to do so, it is harder to imagine them successfully displacing Dun and Bradstreet or Moody's as rating agencies for corporate and government securities.

In short, the impact of e-commerce on intermediation services is more likely to be manifested in a dramatic change in the nature of intermediary services, as well as the identities of intermediaries, rather than in a dramatic reduction in such services. While it is difficult to forecast the changes that will be forthcoming, it seems plausible to suggest that the emergence of new intermediaries is more likely, the greater the changes in the underlying value-added stages of an industry brought about by e-commerce. Equivalently, the emergence and growth of new intermediaries will be more dramatic, the less applicable is old information about supply and demand conditions to the new conditions created by e-commerce.

EVALUATING THE EARLY EXPERIENCE

The relatively limited experience with e-commerce, to date, combined with fragmentary available information on the phenomenon, severely limits one's ability to assess the hypotheses identified in the preceding section. As a result, any conclusions drawn from the e-commerce experience to date are quite tentative.

Expansion of Geographic Markets

The impact of e-commerce on the size of relevant geographic markets will be a function, in part, of how segmented current markets are in locational space, as well as the importance of search costs as a cause of market segmentation. *A priori*, one would expect the law of one price to be violated most frequently in the case of international commerce. For one thing, language and currency differences, as well as differences in product features related to local regulations and cultural attributes, can contribute to geographic price differences being less than fully arbitraged away. For another, tariff and non-tariff barriers may discourage the physical shipment of products across geographic markets. In this context, transportation costs can be seen as discouraging international trade, especially when quick delivery is desired.

As already noted, the growth of e-commerce has been relatively slow outside the US. This apparently reflects, in part, sluggishness on the part of foreign retailers to exploit the capabilities of e-commerce. Hence, in Sweden and Denmark, where Internet penetration is Europe's highest, online shopping is relatively rare.[23] In Canada, on-line sales are only about 5 per cent of US sales (Evans, 1999). An implication is that unless and until on-line markets for products become substantially more developed outside the US, international e-commerce will remain a relatively insignificant channel through which geographically segmented markets will become closely integrated. Even with a more rapid future growth of WWW sites outside of the US, strong local purchasing preferences might act as non-tariff barriers to international e-commerce. Indeed, there is some evidence that such preferences exist. For example, a recent poll found that 70 per cent of regular Internet users in Canada prefer to make on-line purchases at Canadian Web sites. Moreover, the president of a large Canadian on-line bookseller stated that it is an inaccurate perception that the Internet is a global marketplace. Rather, the reality is that consumers like to shop close to home. Among other things, by dealing with local retailers, consumers can buy products using their own currency and also avoid border duties and potentially large shipping costs. Further, attesting to the local nature of commercial Web sites, major commercial portals, such as Alta Vista, highlight retailers that sell in Canadian dollars and ship their products from Canada (Evans, 1999).

The few available case studies of consumer behavior and e-commerce support the notion that, at least at the retail level, buyers continue to prefer to transact with local sellers. An example is the retail brokerage industry, where on-line trading has become a substantial portion of total retail brokerage sales. While there has been a significant shift away from traditional brokerage services to on-line services, the vast majority of retail customers continue to use a local broker (Globerman et al., 2000). Among other reasons, it appears that periodic face-to-face contact with their financial advisors is still an option that many retail customers value. At the same time, it appears that brokerage companies themselves are increasingly buying and selling securities for their customers in an international electronic marketplace. Indeed, a number of mergers and alliances involving national and regional stock exchanges have taken place, apparently motivated, in large measure, by a desire to provide brokerage companies with

Table 4.2   Mergers and alliances among stock exchanges

Exchanges	Features	Announce-ment date
Singapore and Australia	Link trading and settlement system	June 2000
Global Equity Market*	Pass order books across time zones	June 2000
Deutsche Boerse and London	Share trading and regulatory systems specialization by security type	April 2000
Deutsche Boerse and Market XT	Create European broker-dealer to allow US investors to European blue chip stocks	April 2000
Nasdaq and Quebec Government	Co-list securities	April 2000
Nasdaq and Hong Kong	Trade Nasdaq stocks in HK	June 2000
Copenhagen and Stockholm	Integrated trading, clearing and settlement systems	January 1998
Swiss and Tradepoint Financial	Integrated settlement system	May 2000

Note: * A venture involving the New York, Toronto, Tokyo and Hong Kong exchanges and bourses in Paris, Amsterdam, Australia, Brussels, Mexico and Brazil.

Source: Various newspaper reports.

around-the-clock access to markets in which both foreign and domestic securities can be traded. Table 4.2 provides a partial list of some recent mergers and alliances. The amalgamation of national and regional stock exchanges is being facilitated by the international diffusion of relatively standardized technology platforms, as well as a degree of harmonization of international regulations regarding listing procedures and back office practices.

To sum, the experience to date with e-commerce offers grounds for skepticism about how rapidly and dramatically relevant geographic markets for B-to-C transactions will expand. On the other hand, the geographical boundaries for many B-to-B transactions may well expand substantially. While geographical segmentation of wholesale markets is arguably less significant than geographical segmentation of retail markets, the e-brokerage experience suggests that the integration of wholesale markets for financial services is having a substantial impact on the services that brokers are able and willing to offer their retail customers.

Competition and Contestability

On balance, the available evidence suggests that prices on the Internet are lower, on average, than prices for comparable goods purchased through other commercial channels (Schlesinger, 1999). This observation might reflect a greater degree of competition for e-commerce. If so, one would expect that, over time, competition from e-commerce will lead to lower prices for other types of commercial transactions. However, there are also examples of higher prices being associated with e-commerce. For example, it appears that reductions in consumers' search costs lead, in many cases, to increased competition among buyers which, in turn, bids up the prices of particular products.

There are also huge disparities observed across different Web sites for identical, or very similar, products. Hence, the increased competition generated by e-commerce is far from perfect. Moreover, there is some evidence that improved information about consumer buying habits is facilitating increased price discrimination in specific markets. That is, the Internet might actually be facilitating the exercise of market power in specific circumstances. A case study of the retail brokerage industry indicates that new firms have entered segments of the industry in response to opportunities created by e-commerce. Also, prices for services whose costs have been significantly reduced by e-commerce, particularly basic transacting of securities, have dropped substantially. These developments point to an increase in actual, as well as potential, competition within domestic markets for retail brokerage services. Nevertheless, incumbent brokers still enjoy a dominant share of the retail brokerage market, including electronic brokerage.

Table 4.3 reports the estimated number of on-line trades per day, and the market share percentages of the leading US on-line brokers. It should be noted that *de novo* entrants into retail brokerage have enjoyed their greatest success in the US. Nevertheless, of the eight firms listed in the table, only three (E*Trade, Datek and Ameritrade) are new entrants. The other firms are either incumbent discount brokers, or they are divisions of established full-service brokers. The limited amount of actual *de novo* entry suggests that the competitive threat posed by e-commerce has been sufficient to stimulate meaningful competitive responses from incumbents. At issue is whether future technological changes will be sufficiently robust to continue to open up new competitive opportunities for *de novo* and small on-line firms.

As mentioned earlier, a growing concern is that large, established companies might utilize group WWW sites to create and exploit market power through buying and selling on the WWW. An example illustrating this

Table 4.3   Estimated daily trades and market shares

Company	Trades	Market share
Schwab	138250	27.9
E*Trade	65800	13.3
Waterhouse	57800	11.7
Datek	50345	
Fidelity	49981	10.1
Ameritrade	41252	8.3
DLD direct	19062	3.8
Discover	13838	2.8

Source: CBS Market Watch, 'Online trading and online accounts grew substantially in first quarter', 26 April 1999.

potential is the on-line residential real estate market, where Homestore.com has signed exclusive agreements with most local real estate boards to post their listings on Homestore.com's Web site. The National Association of Realtors, in turn, owns part of Homestore.com's residential on-line division. The concern is that this arrangement will make it difficult for realtors who are unaffiliated with The National Association of Realtors to be competitive in on-line real estate sales, since they may be denied access to listing on Homestore's major Web site. To be sure, it is unclear that exclusive commercial arrangements on the WWW are any more (or less) anti-competitive than exclusive commercial arrangements in the world of 'bricks and mortar'. However, commercial arrangements in the world of e-commerce may oblige antitrust authorities to confront competition policy situations for which there is little precedent or experiential knowledge.

Intermediaries

There is little doubt that e-commerce has spawned new sources of intermediation, at the same time as it has altered or eliminated other forms. Most obvious, e-commerce has led to a new form of intermediation known as Web hosting. In this arrangement, major Web portals such as Yahoo provide selling space for unaffiliated retailers. In other cases, major commercial Web sites such as Amazon.com allow unaffiliated retailers to solicit sales on Amazon's site. Amazon also provides billing and transportation services for those retailers. In still other cases, companies have established electronic malls, and smaller companies can lease Web space at those malls.

Both industrial companies and households are using electronic auction services, such as those supplied by Ebay, to buy and sell products. Financial companies are increasingly distributing information through specialized financial Web sites, as well as major portals run by ISPs such as AOL and Microsoft. Indeed, the issue is certainly not whether e-commerce has spawned entire new market intermediaries, it is whether they are strong substitutes for conventional intermediaries. To the extent that e-commerce cannibalizes comparable sales from conventional commercial channels, one would expect there to be perhaps a strong degree of substitution across electronic and non-electronic intermediation. The limited available evidence points to e-commerce largely siphoning off sales that would otherwise have been made through conventional commercial channels. For example, a recent consulting report estimates that just 6 per cent of B-to-C electronic commerce represents new spending.[24]

Obviously, established intermediaries are attempting to make their on-line services complementary to their off-line services. For example, full-service brokerage companies such as Merrill Lynch are bundling on-line trading into a package of services that is made available to customers for a fixed fee. As another example, shopping malls are installing electronic kiosks that allow customers to order goods electronically and then pick those goods up physically at a convenient distribution site in the mall. The point here is that the identities of major electronic commercial intermediaries may change more slowly in the future than in the past, but this should not be taken as an indication that competitive pressures are weakening in this sector of the economy.

CONCLUSIONS

It seems reasonable that a major technological event such as the emergence and growth of e-commerce will change the industrial organization of a wide range of markets; however, the precise nature, magnitude and timing of the changes are difficult to predict. The predominant thinking as expressed in the relevant literature is that the changes will be dramatic, in terms of both the structural attributes of markets, as well as the behavior of market participants. This chapter seeks to identify the underlying linkages between e-commerce and the hypothesized changes in industrial organization. The key linkage is arguably a dramatic reduction in search costs for market participants. It is argued that, while such reductions might well be profound for competition in certain markets, they may be quite modest for others. Moreover, improved information might also, in some cases, facilitate abuse of market dominance and restrictions on competition, rather than increased competition and contestability.

Nevertheless, it seems fair to conclude, based on both theory and the limited available evidence, that the emergence and growth of e-commerce has led to increased competitive innovation on the part of both new and incumbent sellers across a range of markets. Such innovation has been associated with increased product specialization and new product offerings, as well as lower prices. In this regard, simple structural measures such as market share distributions may understate the actual degree of competitive rivalry fostered by e-commerce. Nevertheless, the growth of e-commerce does not obviate the role of competition policy. In particular, fraud, misleading advertising, bait-and-switch and other 'garden variety' abuses against consumers are likely to increase to the extent that hit-and-run entry of dishonest merchants is cheaper in electronic markets than in conventional markets. On the other hand, efficiency arguments in support of practices such as tied selling and below-cost pricing may be more compelling in an e-commerce environment. For example, tied selling may be a convenient way for innovative on-line sellers to internalize the economic rents associated with their commercial innovations, such as by tying the sale of a new, and easily imitated, Internet service to the sale of some other good or service that enjoys some form of intellectual property protection. Below-cost pricing could become a more ubiquitous means of encouraging consumers to try out experience goods. A readier acceptance of potentially abusive competitive behavior in favor of stricter monitoring of access to critical, if not bottleneck, nodes on the WWW may be a beneficial tradeoff for competition policy authorities dealing with the new challenges introduced by e-commerce.

NOTES

1. See 'Is that e-commerce road kill I see', *Business Week*, 27 September 1999, EB 96.
2. Ibid.
3. On-line payments by households are still relatively rare. See 'Internet commerce booms', *The Bellingham Herald*, 22 November 1998, E2.
4. See 'A hard sell online? Guess again', *Business Week*, 12 July 1999, p. 142.
5. See 'Internet retail activity by Canadians', *The Globe and Mail*, 28 January 2000, E5.
6. EDI encompasses B-to-B commercial transactions taking place over electronic communication networks. In its early stages, the bulk of such transacting occurred using private networks as opposed to the WWW.
7. See, 'The click here economy', *Business Week*, 22 June 1998, p. 122.
8. See 'Airlines form online cheap-ticket service', *Seattle Post-Intelligencer*, 30 June 2000, D2.
9. This categorization of transaction costs is discussed in Wigand (1997). A component of search activity is the verification of the claimed attributes of products. Where it is difficult for producers to validate their product claims, markets may be characterized by a 'lemons' problem, and reliable producers may be driven from the market. For a discussion of this phenomenon on the Internet, see Lu (1998).
10. Collectibles, such as rare books, provide an example, see Bensinger (1999).

11. E-commerce also has been suggested to facilitate more complex pricing arrangements that, in turn, could allow producers and consumers to employ multi-part prices that also improve the efficiency of the pricing mechanism.
12. A brief discussion of the distinction between search and experience goods is found in Carleton and Perloff (1990, pp. 596–98).
13. This point is also made in Lu (1998).
14. The emergence of relatively low-cost video conferencing is allowing an increasing number of professionals to have 'face-to-face' consultations with their on-line customers.
15. Supporting this assertion is the observation that the vast amount of on-line medical information, to date, appears to be used by patients primarily to bring ideas and questions to their existing physicians rather than as a source for identifying the services of physicians, see Hafner (1998).
16. A seminal discussion on environmental determinants of opportunistic behavior is found in Williamson (1975).
17. For an enthusiastic statement of how e-commerce spells the end of geography and borders as industrial organizational constructs, see Kobrin (1995).
18. Cooperative behavior is also sometimes characterized as 'conscious parallelism'. For a discussion of this type of behavior see Greer (1992, pp. 394–9)
19. See ibid., pp. 266–9.
20. See 'Now it's your web', *Business Week*, 5 October 1998, pp. 164–78.
21. Suppliers can also search the Internet to see the availability of substitute products and increase prices when the availability of substitutes is limited. See Schlesinger (1999).
22. For a discussion of the growth of on-line retail brokerage trading, see Globerman et al. (2000).
23. See 'The net: Europeans aren't buying', *Business Week*, 6 September 1999, p. 8.
24. See 'Is that e-commerce road kill I see?', *Business Week*, 27 September 1999, EB 96.

REFERENCES

Aeppel, T. (2000), 'A web auctioneer roils the rust belt', *The Wall Street Journal*, 5 January, B1.

Baker, S. and Echikson, W. (2000), 'Europe's Internet bash', *Business Week*, E.Biz, 7 February, EB 40.

Bensinger, K. (1999), 'Untidy shelves: The Internet shakes up rare books', *The Wall Street Journal*, 18 June, W9.

Carleton, D.W. and Perloff, J.M. (1990), *Modern Industrial Organization*, New York: Harper Collins Publishers.

Evans, M. (1999), 'Canadians trail U.S. in e-commerce', *The Globe and Mail*, 20 July, B5.

Globerman, S., Roehl, T. and Standifird, S. (2000), 'Globalization and electronic commerce: Inferences from retail brokering', Western Washington University, mimeo.

Greenberg, L. (1999), 'Canada banks try web to win U.S. customers', *The Wall Street Journal*, 28 October, A18.

Greer, D.F. (1992), *Industrial Organization and Public Policy*, 3rd edn, New York: Macmillan Publishing.

Hafner, K. (1998), 'Can the Internet cure the common cold', *The New York Times*, 9 July, D1.

Klein, B. and Leffler, K. (1981), 'The role of market forces in assuring contractual performance', *Journal of Political Economy*, 89, 615–41.

Kobrin, S. (1995), 'Regional integration in a globally networked economy', *Transnational Corporations*, 4(2), 59–72.

Lu, J. (1998), 'Lemons in cyberspace: A call for middlemen', in E. Bohlin and S.L. Levin (eds) *Telecommunications Transformation: Technology, Strategy and Policy*, Amsterdam: IOS Press, pp. 235–53.

McWilliams, G. (2000), 'PC firms, suppliers plan online exchange', *The Wall Street Journal*, 2 May, B5.

Pfeiffer, H. (1995), *The Diffusion of Electronic Data Interchange*, Heidelberg: Springer-Verlag.

Picot, A., Bortenlager, C. and Rohrl, H. (1997), 'Organization of electronic markets: Contributions from the new institutional economics', *The Information Society*, 13, 107–23.

Schlesinger, J. (1999), 'If e-commerce helps kill inflation, why did prices just spike?', *The Wall Street Journal*, 18 October, P1.

Solomon, S. (1995), 'Staking a claim on the Internet', *Inc Technology*, 16, 87–91.

Wigand, R. (1997), 'Electronic commerce: Definitions, theory and context', *The Information Society*, 13, 1–16.

Williamson, O.E. (1975), *Markets and Hierarchies: Analysis and Antitrust Implications*, New York: The Free Press.

5. The economics of online retail markets

Michael R. Ward

INTRODUCTION

The use of the Internet for the marketing of products and services has grown at an astonishing rate. Measuring this growth, however, has been problematic. Accurate forecasts of the amount of Internet retailing are difficult to make and tend to focus on the US. Indeed, forecasts for the future volume of Internet retailing vary and may seem optimistic (Ward, 2001a). The volume of products purchased in 2000 via the Internet in the US was 30–40 billion United States dollars (USD). Forecasters expect this amount to grow to more than USD100 billion, and possibly exceed USD200 billion, by 2004 (see Table 5.1).

Table 5.1 E-commerce forecasts, 1999–2004 (in USD billion)

Source	Year					
	1999	2000	2001	2002	2003	2004
Emarketer		37.0				>100.0
Forrester	20.6	33.0	52.2	76.3	108.0	184.0
Gartner Group						142.0
Giga Information Group	25.0			152.0		233.0
Lohse, Bellman and Johnson		29.2	46.2	70.5	97.0	

Note: This table compares growth rates for e-commerce expenditure implied by factors affecting growth with previously reported growth rates. The Gartner Group forecast is for North America.

These forecasts represent average annual growth rates of 30–50 per cent. This represents a phenomenal pace of adoption for a new marketing channel, with the most optimistic forecast representing about 3 per cent of US disposable personal income in 2004. Of course, the pace of Internet adoption for electronic commerce (e-commerce) is not likely to slacken much after 2004 and outside the US it could easily increase. Taylor Nelson Sofres Interactive found that among Internet users globally, the US has the highest proportion (58 per cent of population) and also the highest proportion of shoppers (27 per cent). In contrast, only 1 per cent of users currently shop online in Thailand and Turkey (Cyber Atlas, 2000). By all accounts, Internet based retailing is growing rapidly in nearly all but the poorest nations.

This chapter outlines what is different about online retail markets relative to traditional retail markets. Of particular interest is the effect that lower costs of information acquisition by consumers will have on market institutions that develop, and on market efficiency. For example, lower costs of consumer search for product information lead to greater substitutability between homogeneous goods, but perhaps a stronger brand preference for heterogeneous goods. Lower costs of comparing alternatives may result in manufacturers taking on some retailing functions (dis-intermediation) and the new emerging entities providing different retail functions (re-intermediation). Lower delivery costs have led to rapid adoption of online retailing for travel and financial products, while higher delivery costs may be leading to more modest adoption of online retailing for groceries and sporting goods.

MODELING MARKET EFFICIENCY

Generally, more efficient markets generate more total surplus to be allocated among buyers and sellers. Therefore, surplus-seeking consumers and retailers will tend to search out markets that are more economically efficient, giving these markets a comparative advantage over others. Examining those factors that affect the relative efficiency of online and offline markets is a fruitful way of uncovering where, and to what extent, online markets are most likely to succeed. Market efficiency refers to both productive efficiency (how inexpensively the product can be produced) and allocative efficiency (the extent to which the product is assigned to users who value it most). These concepts are usually identified with the economic welfare measures of producer surplus and consumer surplus. Lowering costs while holding prices constant tends to increase producer surplus. Similarly, lowering prices tends to increase consumer surplus. In general, some portion of a firm's reductions in costs is usually passed on in the form of lower prices, leading to both increased consumer surplus and possibly increased producer surplus. Therefore, price reductions are often indicators of increased market efficiency. Note, however, that price reductions could correspond to quality reductions. If so, the decreased willingness to pay could dominate the price reduction, implying a decrease in market efficiency.

For the most part, these concepts can be expressed through the Lerner index. The Lerner index relates a firm's profit-maximizing price–cost margins in an imperfectly competitive market to its demand elasticity as $(P - MC)/P = -1/\eta$. Solving for price, this relation becomes:

$$P = \left(\frac{\eta}{\eta + 1}\right) MC. \qquad (5.1)$$

From (5.1), it can be seen that productive efficiencies that lower marginal costs tend to lower prices and lead to greater market efficiency. Industry changes that increase consumer substitutability between sellers tend to make firm demand more elastic. From (5.1), it can again be seen that this, in turn, lowers the profit-maximizing price and leads to greater allocative efficiency.

Productive efficiencies can arise from reductions in fixed costs and not affect marginal costs. In this case, to a first approximation, they do not affect prices. Therefore, fixed cost reductions increase producer surplus but, because price is unchanged, they do not affect consumer surplus. However, fixed cost reductions can lead to increased consumer surplus indirectly, by making entry of marginal firms viable. Increased consumer substitution toward these new entrants can increase $|\eta|$, which lowers price. Alternatively, entrants may fill an otherwise unserved niche in product space that better aligns consumer preferences with product attributes for some consumers. Either effect would represent increased allocative efficiency.

The product price is only a portion of the total costs incurred by consumers. In addition, consumers may incur non-trivial transactions costs including search, delivery and other financial costs. The full price, P^F, equals the product price, P, plus transactions costs TC. Combining this with (5.1) provides:

$$P^F = \left(\frac{\eta}{\eta + 1}\right) MC + TC. \qquad (5.2)$$

It is possible for the full price paid for merchandise obtained via the Internet to be lower than the full price found through other distribution channels even though the transaction price is higher. This is because some of the consumer activities associated with purchase, such as delivery, can be provided by the firm. A limitation of the above approach is that it does not capture all allocative efficiencies that could flow from better informed consumers. For example, consumers unaware of a difference in two choices may sometimes unwittingly select the item less suited to their preferences. With better product attribute information, consumers would avoid these 'mistakes' and, on average, obtain greater consumer surplus. Differences in either transaction prices or full prices would not capture this form of improved allocative efficiency.

THE INTERNET AND THE EFFICIENCY OF RETAIL MARKETS

Differences in the operations of an online retailer relative to an off-line retailer can affect any number of retailer costs, consumer transactions costs, product information flows, quality assurances, price search effects and distribution channels interactions. For each of these, a body of literature is emerging that evaluates their importance for online markets. In the following section, the ideas and analyses in this literature are reviewed and analyzed.

Retailing Costs

For some product categories, the costs of doing business may simply be lower online than through another distribution channel. For example, it is possible that recording customer identity and consummating payment may be facilitated when the customer directly inputs this information rather than when taken in verbally, as is the case when a mail order catalog retailer receives a telephone order. For intangible goods, such as software, music or information goods, delivery costs can be greatly reduced by using the Internet. In these cases, both nominal and full prices will tend to be lower online.

Many physical products that are shipped across state lines may also avoid sales taxes. For some product categories, such as consumer electronics, this can represent substantial savings. Lukas (1999) and McClure (1999) discuss the distortions that the current tax regime creates. Goolsbee (2000) finds empirical evidence that the tax advantage of online commerce does affect consumers' purchase decisions. He notes that removing this distortion would not greatly increase tax revenue because much of the commerce conducted online could migrate to mail order. Goolsbee (1999) further investigates consumer responses to differential tax treatment to find that newer and less experienced Internet users are less sensitive to tax effects. Still, Goolsbee (2001), and Goolsbee and Zittrain (1999) analyze the sources of state revenues and conclude that Internet-related commerce will not have a substantial affect for the foreseeable future.

Online retailers incur fixed costs in the development of their web presence. However, online-only retailers need not incur the fixed costs associated with retail outlets in different geographic markets. Online retailers may be better able to manage inventories because they can better incorporate retail and wholesale functions. Therefore, a distribution center providing inventory for online retailers would be larger than for comparable off-line retailers. However, while off-line and hybrid retailers (distributing through both online and in-store channels) may serve only a small geographic market, Internet only stores can serve all customers covered by their delivery company. Because of this increased reach, the larger facility used by online retailers provides services for a disproportionately larger number of customers. The average fixed cost per unit sold can often be much lower for online retailers than for off-line retailers.

Most models of monopolistic competition yield price reductions or increased product differentiation with lower average fixed costs. In models of homogeneous goods with fixed costs and free entry, as in Chamberlain (1933), when average fixed costs fall, entry occurs and prices tend to fall. In differentiated goods models, as in Spence (1976) and Dixit and Stiglitz (1977), when average fixed costs fall, entry also occurs, but now entrants tend to more fully fill product space leading to greater product variety and possibly lower prices. It is likely that the greater reach of retailers through the web will allow for more firms to be viable at a smaller scale. Thus it is possible that online marketers, like catalog marketers, are able to extend their geographic reach while simultaneously narrowing their focus on niche product markets.

Transactions Costs

Online retailers typically deliver the order to the consumer. For online purchases that replace off-line purchases, like consumer electronics, this represents a shift of some transactions costs from the consumer to the marginal costs of the retailer. The total effect on the full price, P^F, could be either to increase it or decrease it and will depend on retail strategy and particular consumer shopping patterns. As mentioned above, for products and services that can be delivered electronically, such as software, music and information goods, the transactions costs and, thus, the full price should fall. However, for physically large products, such as furniture or construction products, the delivery costs are likely to be so much higher when provided by the retailer than by the consumer, that an online market will only cater to those consumers with extremely high delivery costs, such as the homebound. The cost of market-provided delivery relative to consumer-provided delivery for a good may be a good predictor of the feasibility of Internet-based retailing.

Product delivery method is independent of the market channel for certain goods. For example, retailers often deliver bulky items such as white goods (e.g. refrigerators and dishwashers) to consumers' homes even in the off-line market. In this case, however, the consumer often will wish to observe and select the product in a showroom prior to purchase. Even though there are no additional delivery costs, Internet retailing for appliances has not developed greatly. For similar reasons, customers of online grocery retailers may not observe lower delivery costs. Internet retailers, especially those that specialize in non-perishable items, may supply only a portion of the items on the grocery list. Their customers will not see a decline in their own visits to an off-line store comparable to the increase in delivery costs from the Internet retailer. However, some consumers are already splitting their food purchases between stores for pantry items and higher end grocers for perishable items (Morganosky, 1987). Likewise, some consumers may reduce their store visits, and their concomitant transactions costs, by using the Internet to purchase groceries.

Online retailers usually gather the specific items requested by the customer. In some cases, like prescription pharmacy, retailers already physically fill the shopping basket of items to be purchased. In other cases, like books, music or groceries, the Internet retailer is now incurring a cost previously borne by the customer. When Internet retailers incur these costs, customers no longer do so. This will tend to lead to a higher transaction price, but not necessarily a higher full price. In some cases, like music, automation of this process may favor Internet retailers. The amount of the transactions cost savings depends on the opportunity cost of the consumer's time with larger benefits accruing to those who value their time more, such as upper income and dual-career households. In others, as with groceries, gathering items can be intricately linked with the consumer choice decisions, meaning that automation may come at the sacrifice of traditional aids to shopping.

Product Information

The vast array of consumer products available requires mechanisms to sort through the possible choices. Before price comparisons are made, consumers often reduce the relevant choices to a manageable consideration set. However, the process of winnowing down the choices involves a number of mechanisms. For example, store retailers tend to offer a selection of substitute products close to each other to facilitate these comparisons. Likewise, brand names and shopping malls provide mechanisms for reducing consumers' costs associated with product selection (Pashigian and Bowen, 1994). The Internet may provide alternative comparison mechanisms that will lead to radically different transactions cost lowering market institutions.

Indeed, e-commerce is driving researchers to rethink the role of intermediators (Gellman, 1996). Online retailers may not have as substantial a comparative advantage in facilitating product comparisons over in-store retailers as once believed. For many product attributes, consumers can easily obtain product information from manufacturer web sites. Shopbots may emerge as a viable mechanism online for the provision of this retailing function (Brynjolfsson and Smith, 2000). If so, exchange may increasingly be accomplished without the explicit intermediation of retailers. Other product attributes may not be so easily conveyed over the Internet. For these, online retailers may provide a signal to consumers of the features of products they carry. For example, shopping at the limited.com and BrooksBrothers.com may be the simplest way for consumers to assure themselves of certain styles and qualities. Direct manufacturer business-to-consumer (B2C) sales require transactional inventory infrastructure that many manufacturers currently do not possess. Manufacturers may choose to outsource these intermediary functions to behind-the-scenes specialist firms (Chircu and Kauffman, 2000).

Internet auctions represent a growing mechanism for consumers to obtain information about product availability (Lucking-Reiley, 2000). Using auctions sites, consumers can easily find unique items that match their preferences in product attributes. In this case, the Internet has probably reduced total transactions costs sufficiently that many items of modest pecuniary value are now exchanged that otherwise might have been discarded. Moreover, Internet auctions provide a record of product and transaction information that can be exploited by researchers. For example, Lucking-Reiley et al. (2000) measure the value of seller reputation in terms of price premiums, and List and Lucking-Reiley (2000) test the dominance of one auction mechanism over another.

The Internet may allow for more accurate segmentation and target marketing than previous technologies. Target marketing allows retailers to provide specific product information to those consumers who are most likely to derive benefit from their product. For example, a ski outfitter may want to send more direct mailings to consumers in Vermont than those in Hawaii. Target marketing, in this way, lowers the average cost of selling, allowing more surplus to be captured by consumers and retailers. Online target marketing can use consumer information more immediately and at a much finer level; for example, at the household level versus geographic location. Therefore, marketers who have access to and use this information are at a comparative advantage relative to those who do not.

Use of consumer level data by online marketers often raises privacy concerns. It is assumed that most consumers do not wish for strangers to know too much about their personal life, especially in relation to their health and finances. Policymakers are currently grappling with the appropriate tradeoffs between consumer privacy and increased consumer and producer surplus from improved access to consumer information (Bennett and Grant, 1999). While measuring the value of consumer privacy may prove to be intractable, some estimates of its costs are obtainable. A study sponsored by the Direct Marketing Association (Turner, 2001) claims that a proposed restriction limiting marketer access to, and use of, consumer information, both online and off-line, would raise a marketer's costs by 3.5 per cent to 11 per cent on average. In addition to privacy concerns, some fear that access to consumer information may allow marketers to price discriminate. Online retailers may be able to infer that they have more market power over an individual consumer from a customer's profile. If so, they may offer products at higher prices to these customers than they do to other consumers (Clemons et al., 2000). It should be noted that the welfare implications of increased ability to price discriminate are uncertain (Varian, 1985). Moreover, even with more price discrimination, average price–cost margins could still be less online if consumers are more price sensitive on average.

Quality Assurance

While the Internet may help a consumer to find products that claim to have his or her desired attributes, not all claims are to be believed. Some attributes, primarily quality related, are not directly verifiable at the time of purchase, requiring some form of quality assurance. This could undermine the viability of a market (Akerlof, 1970). Assurance from the expertise or certification of third parties in off-line markets comes from such institutions as *Consumer Reports*, and the *Good Housekeeping* and Underwriters Laboratories Seals of Approval. The counterparts that have developed in online markets include Bizrate.com and Epinions.com. Demonstrations of product quality as a means of assuring quality are thought to be a cause of certain vertical restrictions (Telser, 1960). The selling of a high-price, high-quality product may best be done through point-of-sale demonstration services rendered by the retailer. Manufacturers have encouraged these demonstrations by providing retailers particularly high product margins to cover the extra costs. A problem arises when some retailers forgo the services, but instead free-ride off those services provided by others. To discourage free-riding, a manufacturer may wish to place contractual restrictions on what retailers must offer. Carlton and Chevalier (2001) find evidence that the restrictions found in off-line markets carry forward into online markets.

A purpose of brand advertising is to provide quality assurance (Nelson, 1970, 1974; Nichols, 1998). For products with experience attributes, where actual quality can only be verified on experiencing the good, high-quality producers can signal their product's quality through expensive advertising. It is difficult for low-quality producers to profitably duplicate this strategy because consumers eventually discover their product's actual quality and forgo the repeat purchases needed to cover the advertising expense. This theory predicts that the large advertising expenditures need not convey detailed product information. Explicit brand advertising has developed online primarily as banner ads. However, simply posting product information to a retailer's web site can be considered advertising. Indeed, retailers claim that the primary use of web sites is for marketing and product information dissemination purposes (Direct Marketing Association, 2000).

Interesting questions remain about the substitutability of brand advertising for experience goods and the use of product feature information for search goods. Some answers may be found in online markets. The richness of information flow on the Internet may provide a means of conveying enough product information so that some experience attributes are rendered search attributes. For example, vacation destination hotels rely on branding because the quality and features of the hotel are often unknown prior to the vacation. However, with a website, a hotel may be able to provide photographic evidence of many of the more important hotel attributes. If so, the Internet may cause national branding to become less important and may increase the variety of the offerings of hotel accommodation. Ward and Lee (2000) find that, as users gain experience with the Internet, they become more proficient at searching for products and less reliant on brand advertising. This suggests that consumer search and retailer advertising may be substitute mechanisms for providing product information.

Price Search Effects

Since Stigler (1961), economists have believed that lower search costs make markets more competitive and increase allocative efficiency. With lower search costs, consumers are likely to become more informed about alternative retailers for a specific product and are more likely to purchase from the lower priced retailers. Essentially, better information increases the substitutability between retailers, causing firm-level demand to be more elastic and leading producers to price more closely to marginal cost. A growing number of studies have examined the pricing in online markets. Clemons et al. (2000) find that online travel agents return different quotes for air travel for the same request between 2.2 per cent and 28 per cent of the time and conclude that the market is not perfectly competitive. Bailey (1998) finds that online booksellers are more expensive, while Brynjolfsson and Smith (2000) find that music and book prices are both lower online, suggesting that online markets are more competitive. Scott Morton et al. (2000) find that prices for automobiles are, on average, 2 per cent or USD450 lower when consumers search online. Moreover, they break down this price reduction into portions related to lower priced dealer selection and increased bargaining power on the part of consumers. They claim that the lower price dispersion is consistent with markets being better disciplined through more consumer information. Lee (2000) finds lower online prices for prescription drugs, Brown and Goolsbee (2000) find lower online prices for insurance, but Ward (2001b) finds higher online prices for groceries.

These same studies, and others, often attempt to draw inferences about the degree of competition from patterns of price dispersion. Bailey (1998) and Brynjolfsson and Smith (2000) infer from less disperse online book prices that these markets are more competitive. Png et al. (2001) have recently criticized this conclusion by noting that price could be lower online even though margins are higher if costs are sufficiently lower and that lower price dispersion is consistent with both competition and collusion. Instead, they propose tests that associate prices with measures of differences in market institutions such as measures of lower search costs. Indeed, Kauffman and Wood (2000) find evidence that the Internet increases firms' ability to tacitly collude. Ward (2001b) argues that comparisons of price levels could be misleading since some of the non-price transactions costs that are borne by the consumer in off-line markets may be borne by the retailer in online markets. Instead, findings that link price dispersion to factors related to the form of competition allow for a more powerful test of the degree of competition.

Only Brown and Goolsbee (2000) measure price dispersion based on units sold. Most studies take posted prices as reported by retailers as the unit of observation without information about the number of units sold at a given price. Better informed consumers online could skew purchases toward lower priced retailers relative to off-line markets. At the same time lower entry barriers online could make entry as a high price but low volume retailer profitable. Thus it may be possible to measure more dispersion in price offerings weighted by retailers online even though there is less price dispersion in actual prices paid weighted by units sold. Dispersion could also reflect price discrimination caused by unilateral strategic pricing. Varian (1980) outlines a model in which firms charge high regular prices to loyal customers and hold periodic sales that provide disloyal customers, who maintain private inventories, with lower average prices. Hoskens and Rieffen (1999) and Lach (2000) find evidence of this behavior for off-line retail markets. Lee (2000) and Ward (2001b) find price volatility patterns consistent with diminished ability to engage in this form of price discrimination online. Still, Baylis and Perloff (2000) find evidence of market segmentation in which some retailers charge low prices and provide good service targeted toward the informed consumers, while others charge higher prices and provide worse service targeted toward 'uninformed' consumers.

Other studies attempt to measure substitutability directly. These suggest that increased product information may not always translate into increased consumer price sensitivity online (Lynch and Ariely, 2000; Degeratu et al., 1999; Shankar et al., 1999). These results tend to contradict the idea that more information in online markets will increase competition and lead to prices closer to marginal costs. A possible explanation is that consumers are receiving more information both about price and other product attributes. At the same time that the price information facilitates brand switching, information about product attributes could better identify the choice that best satisfies a consumer's demand. They might have been more price sensitive when they did not know which choice was best, but now that they do know, a larger price difference is required to induce them to switch.

Effects on Other Channels

Online retailing is likely to affect traditional markets. Merrill Lynch famously slashed its brokerage fees in response to increased competition from online brokerages. In this way, increased market efficiency online could induce increased efficiency in off-line markets. The magnitude to which this spillover will occur depends on cross-channel substitutability. Some evidence suggests that cross-channel substitutability is high. As discussed above, Goolsbee (2000) provides some evidence of substitutability between off-line and online channels. For a number of product categories, Ward (2001a) investigates the substitutability between online, in-store and direct mail channels from the perspective of transferable human capital investments and finds more substitutability between the online and direct mail channels than the online and in-store channels. Ward and Morganosky (2000) investigate linkages between the product search channel and the product purchase channel and find greater inter-channel linkages from online information gathering to in-store and direct mail purchases than any other unaligned search–purchase combination.

CONCLUSION

Online retailing is still new. It is probably most advanced in the US, but only represents about 1 per cent of all retail sales currently. This is expected to continue to grow at a tremendous rate. Within five years, online retailing could represent 5–10 per cent of all retail sales in the more developed countries. Moreover, at this point online retailing could have more profound effects on off-line retailing. Studying what is happening today will help us to understand what types of effects are likely to occur. Generally, online retailing is likely to make markets more efficient. For some products, the costs of marketing products to consumers online will be lower. Products like software, travel arrangements and financial services benefit from lower delivery costs online so that the full price will fall for most consumers and online marketing will largely replace other channels. For other products, the transaction price may be higher, but it will include services not currently offered that are favored by at least some segments of the market. For these segments, therefore, the full price is lower than the opportunity cost. Finally, the costs of marketing some products online, such as building materials, will probably remain prohibitive for the foreseeable future.

Perhaps the more interesting changes brought about by e-commerce stem from the amount and type of information available to both buyers and sellers. Gathering product information and making product comparisons will continue to become easier for consumers. This is likely to affect both their product search behavior and firms' product advertising behavior. These changes in consumer information should affect their decisions and even their decision processes. Firms, faced with better informed consumers, but armed with more consumer information themselves, will abandon some marketing strategies and develop new ones. Economic theory can help facilitate our understanding of which strategies are more effective under different information structures. Already, a growing literature has attempted to apply this theory to specific online markets. Undoubtedly, online marketing will provide many settings for understanding consumer use of information and firms' strategic responses.

ACKNOWLEDGEMENTS

I would like to thank Mike Lee and Michelle Morganosky for their comments and suggestions.

REFERENCES

Akerlof, G. (1970), 'The market for lemons: Quality uncertainty and the market mechanism', *Quarterly Journal of Economics*, 84, 488–500.

Bailey, J.B. (1998), 'Internet price discrimination: Self-regulation, public policy, and global electronic commerce', working paper available at http://www.rhsmith.umd.edu/tbpp/jbailey/pub/discrimination.pdf.

Baylis, K. and Perloff, J.M. (2000), 'Price dispersion on the Internet: Good firms and bad firms', working paper at http://are.berkeley.edu/~perloff/PDF/internet.pdf.

Bennett, C. and Grant, R. (1999), *Visions of Privacy: Policy Choices for the Digital Age*, Toronto, University of Toronto Press.

Brown, J.R. and Goolsbee, A. (2000), 'Does the Internet make markets more competitive?: Evidence from the life insurance industry', working paper, at http://gsbwww.uchicago.edu/fac/austan.goolsbee/research/insure.pdf.

Brynjolfsson, E. and Smith, M. (2000), 'The great equalizer? Consumer choice behavior at Internet shopbots', at http://ebusiness.mit.edu/papers/tge/tge.pdf.

Carlton, D.W. and Chevalier, J.A. (2001), 'Free riding and sales strategies for the Internet', NBER working paper 8067 at http://www.nber.org/papers/w8067.

Chamberlain, E.H. (1933), *The Theory of Monopolistic Competition*, Cambridge, MA, Harvard University Press.

Chircu, A. and Kauffman, R. (2000), 'Reintermediation strategies in business-to-business electronic commerce', *International Journal of Electronic Commerce*, 4, 7–42.

Clemons, E.K., Hann I-H. and Hitt, L. (2000), 'The nature of competition in electronic markets: An empirical investigation of on-line travel agent offerings', working paper at http://grace.wharton.upenn.edu/~lhitt/etravel.pdf.

Cyber Atlas (2000), 'Worldwide e-commerce penetration stands at 10 percent', at http://cyberatlas.internet.com/markets/retailing/article0,1323,6061_423751,00.html.

Degeratu, A., Rangaswamy, A. and Wu, J. (1999), 'Consumer choice behavior in online and traditional supermarkets: The effects of brand name, price and other search attributes', Penn State eBusiness Research Center Working Paper 03-1999 at http://www.ebrc.psu.edu/papers/abstract/03_1999.html.

Direct Marketing Association (2000), 'The DMA's state of the interactive e-commerce marketing industry report 2000: Emerging trends and business practices', at http://www.the_dma.org/library/publications/interactiveecommerce.shtml.

Dixit, A.K. and Stiglitz, J.E. (1977), 'Monopolistic competition and the optimum product diversity', *American Economic Review*, 67, 297–308.

Gellman, R. (1996), 'Disintermediation and the Internet', *Government Information Quarterly*, 13, 1–10.

Goolsbee, A. (1999), 'Internet commerce, tax sensitivity, and the generation gap,' in J. Poterba (ed.) *Tax Policy and the Economy*, 14, Cambridge, MA, MIT Press.

Goolsbee, A. (2000), 'In a world without borders: The impact of taxes on Internet commerce', *Quarterly Journal of Economics*, 115, 561–76.

Goolsbee, A. (2001), 'The implications of electronic commerce for fiscal policy (and vice versa)', *Journal of Economics Perspectives*, 15, 13–24.

Goolsbee, A. and Zittrain, J. (1999), 'Evaluating the costs and benefits of taxing Internet commerce', *National Tax Journal*, 52, 413–28.

Hoskins, D. and Rieffen, D. (1999), 'Pricing behavior of multiproduct retailers', mimeo.

Kauffman, R.J. and Wood, C.A. (2000), 'Analyzing competition and collusion strategies in electronic marketplaces with information asymmetry', working paper, available at http://misrc.umn.edu/wpaper/WorkingPapers/CompCollStrat.pdf.

Lach, S. (2000), 'Existence and persistence of price dispersion: An empirical analysis', mimeo.

Lee, M.J. (2000), 'A comparative analysis of pharmaceutical pricing', mimeo, available at http://www.students.uiuc.edu/~mjlee/compare1.pdf.

List, J. and Lucking-Reiley, D. (2000), 'Demand reduction in multi-unit auctions: Evidence from a sportscard field experiment', *American Economic Review*, 90, 961–72.

Lucking-Reiley, D. (2000), 'Auctions on the Internet: What's being auctioned, and how?', *Journal of Industrial Economics*, 48, 227–52.

Lucking-Reiley, D., Bryan, D., Prasad, N. and Reeves, D. (2000), 'Pennies from eBay: The determinants of price in online auctions', available at http://www.vanderbilt.edu/econ/reiley/papers/PenniesFromEBay.pdf.

Lukas, A. (1999), 'Tax bytes: A primer on the taxation of electronic commerce', Cato Institute, Trade Policy Analysis, No. 9, available at http://www.freetrade.org/pubs/pas/tpa_009es.html.

Lynch, J.G. and Ariely, D. (2000), 'Wine on-line: Search costs and competition on price, quality and distribution', *Marketing Science* 19, 83–103.

McLure, C.E. Jr (1999), 'Electronic commerce and the state retail sales tax: A challenge to American federalism', *International Tax and Public Finance*, 6, 193–224.

Morganosky, M.A. (1987), 'Format change in US grocery retailing', *International Journal of Retail and Distribution Management*, 25, 211–18.

Nelson, P. (1970), 'Information and consumer behavior', *Journal of Political Economy*, 78, 311–29.

Nelson, P. (1974), 'Advertising as information', *Journal of Political Economy*, 82, 729–54.

Nichols, M. (1998), 'Advertising and quality in the US market for automobiles', *Southern Economic Journal*, 64, 922–39.

Pashigian, B.P. and Bowen, B. (1994), 'The rising cost of time of females, the growth of national brands, and the supply of retail services', *Economic Inquiry*, 32, 33–65.

Png, I., Lee, S.-Y.T. and Yan, C. (2001), 'The competitiveness of on-line vis-a-vis conventional retailing', at http://e_commerce.mit.edu/papers/ERF/ERF104.pdf.

Scott Morton, F., Zettlemeyer, F. and Silva Risso, J. (2000), 'Internet car retailing', NBER Working Paper W7961 at http://papers.nber.org/papers/W7961.pdf.

Shankar, V., Rangaswamy, A. and Pusatari, M. (1999), 'The online medium and customer price sensitivity', Penn State eBusiness Research Center Working Paper 04-1999 at http://www.ebrc.psu.edu/papers/abstract/04_1999.html.

Spence, M.A. (1976), 'Product selection, fixed costs, and monopolistic competition', *Review of Economic Studies*, 43, 217–36.

Stigler, G.J. (1961), 'The economics of information', *Journal of Political Economy*, 69, 213–25.

Telser, L.G. (1960), 'Why should manufacturers want fair trade?', *Journal of Law and Economics*, 3, 86–105.

Turner, M.A. (2001), 'The impact of data restrictions on consumer distance shopping', Information Services Executive Council, Direct Marketing Association, http://www.the_dma.org/isec/9.pdf.

Varian, H.R. (1980), 'A model of sales', *American Economic Review*, 70, 651–9.

Varian, H.R. (1985), 'Price discrimination and social welfare', *American Economic Review*, 75, 870–75.

Ward, M.R. (2001a), 'On forecasting the demand for e-commerce', in D.G. Loomis and L.D. Taylor (eds) *Forecasting the Internet: Understanding the Explosive Growth of Data Communications*, Boston, Kluwer Academic Publishers.

Ward, M.R. (2001b), 'Will online shopping compete more with traditional retailing or catalog shopping?', *Netnomics*, 3(2), 103–17.

Ward, M.R. and Lee, M.J. (2000), 'Internet shopping, consumer search and product branding', *Journal of Product and Brand Management*, 9.

Ward, M.R. and Morganosky, M. (2000), 'Online consumer search and purchase in a multiple channel environment', working paper, available at http://ux6.cso.uiuc.edu/~ward1/cannibal.PDF.

6. Regulation for Internet-mediated communication and commerce

Robert M. Frieden

INTRODUCTION

This chapter will address the problems and opportunities presented by proliferating and diversifying services provided via the Internet. After reviewing the different types of Internet-mediated services and the pre-existing regulatory models that apply, the chapter will address whether and how a single legal and regulatory foundation exists for addressing the manifold public policy issues raised by Internet-delivered communications and electronic commerce. The chapter concludes with suggestions on what governments should undertake to promote widespread and robust use of telecommunication and information processing infrastructure as conduits for services that promote education, rural development and commerce.

A MULTI-FACETED INTERNET

The Internet means different things to different people. On a technological level it constitutes a 'network of networks' in the sense that Internet Service Providers (ISPs) link their individual networks into an integrated network of networks. ISPs provide consumers with seamless access to most of the individual networks that comprise what we call the Internet, often with a contract covering only the first or last of many network connections. The packet-switched nature of the Internet coupled with switching and routing protocols, provides robust and diverse network access without each ISP having to negotiate interconnection terms with every other operator. Telecommunications carriers achieve similar connectivity with greater effort and specificity through the one-by-one accumulation of operating agreements.

Internet users benefit from the technological ease in switching and routing traffic, but such seamlessness generates a host of legal and regulatory problems. For example, the lack of contract privity between each and every ISP raises liability questions when an ISP inadvertently provides a conduit for a criminal transaction, such as the transmission of obscenity, serving as the delivery mechanism for securities fraud, or providing the forum for predatory, libelous or other illegal behavior. The legal and regulatory models created for telecommunication carriers provide near absolute exculpation. As neutral and transparent common carriers, telecommunication service providers avoid liability or culpability even when serving as the conduit for the commission of a crime.

ISPs do not operate as common carriers. They benefit by not incurring the duties of common carriers: to provide service to any and all users in a particular geographical region without discrimination. But they suffer by potentially incurring liability for the adverse impact and consequences of the content they carry. In the absence of an Internet-specific legal and regulatory model, courts and other decision makers have applied and extrapolated from pre-existing models developed for more established, conventional media. By analogy the Internet becomes the functional equivalent to a common carrier, but only in some instances. Under other circumstances, ISPs do bear some, possibly substantial, responsibility for the content they carry. For example, some courts considering the liability of ISPs for serving as the disseminator of libelous, indecent or copyright violating content have analogized the Internet to pre-existing models applicable to publishers, distributors and sellers of print material. The use of analogies and extrapolations does not work well for two primary reasons. First, the absence of a single, universal jurisprudential and regulatory model means that Internet operators lack certainty as to what standard of care and content scrutiny they must assert. Operators may 'chill' speech by overzealousness, but others may incur substantial financial liability should they fail to self-regulate with appropriate severity and comprehensiveness. Second, the proliferation of services, features and functions available via the Internet defies compartmentalization into traditional print, broadcast and common carrier models. An ISP does not necessarily become a newspaper publisher simply because it delivers an electronic edition, nor does it become a radio station simply by simulcasting the content.

ALTERNATIVE REGULATORY MODELS

The versatility of the Internet defies compartmentalization and an easy extension of existing regulation. First, one would be hard pressed to determine which of several regulatory models to apply. Internet mediation provides faster, better, cheaper and more convenient applications of pre-existing media like books, newspapers, magazines, radio, television, cinema, telephones, facsimiles, and postal mail. Second, pre-existing regulatory model candidates include those outside the communications media environment. The Internet can also serve as a medium for various other types of transactions involving banking, securities, adult-oriented products and services and other types of commerce subject to sector-specific regulation.

The subject of Internet regulation has triggered vigorous and sometimes heated debate. At one end of the continuum, libertarian advocates see no need for any government intrusion. This viewpoint extrapolates from the largely unregulated information technology environment where a mostly unfettered marketplace has rewarded entrepreneurship, creativity and the 'killer application' far better than government handicappers. Advocates for an unregulated and untaxed Internet believe that the absence of government involvement has directly contributed to the enormous success achieved in such a short time.

The absence of barriers to market entry has made it possible for speedy debut of new products and services. Further, the relative ease of access to capital, optimism that the Internet can change everything and the perception that one can invest in the next IBM accounts for the success of Internet equipment and service providers. But one cannot discount the favorable impact created by the absence of drag that government regulation could have imposed. Builders of the next best thing have not needed government authorization to enter the marketplace. Market forces, with all the excesses and potential for good and bad, have largely determined winners and losers. Cisco has captured the dominant market share of Internet routers even in the absence of government-created equipment standards. Microsoft's Windows computer operating system serves over 90 per cent of the world's personal computers, because of marketplace forces, including the impact of increasing consumer benefits (through positive network externalities), accruing when more computers use the same operating system and compatible software. Industry leaders like America On-line, Amazon.com, Yahoo and Expedia have acquired market share by building the better mousetrap, and not by winning a government-administered beauty contest.

On the other hand, one cannot say that the Internet operates regulation free. Courts have not exempted from civil or criminal liability individuals who commit torts or crimes using the Internet as a medium to carry out the activity. Individuals have been convicted of using the Internet as an instrument for crime in such diverse areas as obscenity, gambling, securities fraud, money laundering and many types of consumer fraud. As a network of networks the Internet makes it difficult to identify the physical location where a transaction has taken place. This lack of a physical place as we customarily understand it, makes it difficult to assert jurisdiction over illegal transactions and to have effective enforcement of laws across borders, even when governments have agreed to cooperate and enforce the laws of another nation. The ambiguity of place also makes it easy for criminals to operate, because they can more easily mask their activity, exploit more liberal laws and privacy protection and obscure their identity and the nature of their activity. The scope of the problem runs the gamut from individual illegal activities – for example, gambling, importation of alcohol, easier access to pornography without shame – to group activities that may have a substantial impact on national finances – for example, money laundering, tax avoidance and consumer fraud.

POSSIBLE MARKET FAILURE IN CYBERSPACE

Market failure in the Internet may occur, because some stakeholders predictably will object to the outcomes generated by market forces. Another type of failure can occur – the inability of conventional legal and regulatory regimes to enforce pre-existing laws and to detect violations. In some instances these types of failures combine in ways that galvanize pro-regulatory forces. For example, most nations make it a crime to distribute obscenity and require purveyors of indecent, but not obscene, materials to restrict access to adults. The Internet provides more convenient opportunities for access to pornography and often children have access to such material in the absence of affirmative efforts by parents or guardians to filter out such content, or to monitor a child's Internet consumption. One could argue that market failure occurs when society, including children, has greater opportunities to indulge in its appetite for restricted and clandestinely desired materials. In a non-Internet mediated environment, law and policy makers could crack down on pornography sellers for selling obscene materials, or for selling materials to minors. Internet mediation makes it far more difficult for government officials to perform this function, thereby shifting the policing burden to parents and guardians, or software filters.

Internet mediation provides faster, better, cheaper and more convenient opportunities for accessing and participating in a variety of troubling endeavors. Users have largely undetected access to information about subversive activities including bomb making. Individuals can make terrorist threats and invade another's privacy. By encrypting Internet traffic, criminals can more easily launder money and engage in illegal activity. The ability for one user to distribute content to virtually all other Internet users exacerbates concerns over wrongful expropriation of intellectual property. On this point, the largely unregulated nature of the Internet makes it possible for cyber-squatters to register domain names, such as World Wide Web addresses like www.cnn.com, solely with an eye toward extorting money.

A libertarian philosophy, predisposed against regulation, coupled with concerns about freedom of expression in an open marketplace of ideas militates against government intervention. Yet real financial, emotional and physical harm can and does result from Internet-mediated transactions. At the very least criminals; that is, violators of individual rights, should not escape liability simply because they perpetrated the crime via the Internet.

TRADITIONAL REGULATORY CONSTRUCTS

When considering new technology, various stakeholders, including jurists, regulators and legislators, have an inclination to extend or extrapolate from pre-existing models. Some observers might consider this a fruitless and a wrong-headed view that 'there's nothing new under the sun'. Others approve of a process that strives for continuity and consistency. For example, the concept of free speech protected in the United States by the First Amendment to the Constitution has survived the challenges presented by different and new media. A significant tension exists between advocates for extrapolation and extension of pre-existing models, and proponents of a new model that reflects substantially changed circumstances. Courts often use legacy models to conserve judicial resources and to display restraint based on the view that they interpret law rather than make policy. Regulators may have incentives to extend the wingspan of their jurisdiction and responsibilities. On the other hand, extension of pre-existing models may apply ill-fitting classifications and impose unnecessary restrictions:

> New technologies, while perhaps similar in appearance or in functionality, should not be stuffed into what may be ill-fitting regulatory categories in the name of regulation. Rather, . . . [regulators] should continue the approach of studying new technologies and only stepping in where the purpose for which the Commission was created . . . demands it. (Oxman, 1999)[1]

The rights and responsibilities historically vested in common carriers tempered their market power in exchange for reduced liability or insulation from commercial and personal damages caused by the content carried.[2] Historically providers of neutral and transparent conduits did not have to monitor the content carried,[3] nor could they typically refuse access to their bottleneck facilities[4] on the basis of content.[5] Arguably ISPs operate like common carriers, at least insofar as their carriage of content generated by others. Furthermore, one could assert that the common carrier insulation from liability supports the development of ubiquitous ISP infrastructure and in turn universal access to information services. On the other hand, common carriage historically has applied exclusively to public utilities and other providers of essential services. Policy makers have not yet deemed Internet access so essential as to place it in the same category as 'Plain Old Telephone Service' as opposed to other desirable, non-common carrier services like cable television.[6]

Additionally, recent developments in the interpretation of what constitutes common carriage does not support extending the classification to ISPs, or using it as a vehicle to bolster public policy support for universal Internet service. The dichotomy between common carriers and private carriers has grown murky, because of legislative and regulatory tinkering with the common carrier model (Frieden, 1995); technological innovations; a growing body of cases articulating robust speaker rights of common carriers; and court cases imposing quasi-common carrier obligations on private carriers – for example, the duty of cable television operators to carry broadcast television signals – and quasi-publisher duties on common carriers – for example, the duty to inquire and disclose whether content is obscene or indecent.[7]

Extension of the common carrier model appears difficult now that common carriers can avoid many of the traditional requirements, and non-common carriers have acquired some of the insulation from liability previously available only to common carriers. In the United States, the recent omnibus revision to the organic telecommunications law has made the sense of what constitutes common carriage even murkier.[8] The Telecommunications Act of 1996 requires the application of common carriage classification to commercial providers of telecommunication services,[9] but authorizes the Federal Communications Commission (FCC) to abandon virtually all regulatory requirements on any common carrier if circumstances favor such deregulation.[10] On the other hand, the Act provides ISPs with a quasi-common carrier exemption from liability for the carriage of material, like obscenity and copyright violations, if the operator had no knowledge that it carried the offending content.[11]

The broadcast regulatory model imposes public service responsibilities on operators based on their use of the public radio spectrum. Broadcasters operate as public trustees and must serve the public interest as a fair return in exchange for free use of the spectrum. The broadcast model permits government to impose significant behavior and content regulations, some of which reduce the scope of broadcasters' opportunities for self-expression. This model would lend itself to imposing desirable universal service and public access responsibilities on ISPs. However, the lack of spectrum usage by ISPs eliminates a linkage with pubic service responsibilities. Imposing such restrictions, without the use of spectrum or another compelling justification, would appear to impose unfair restrictions on ISP freedom of expression.

The cable television regulatory model operates a hybrid that blends free expression rights with public interest burdens. In the United States the FCC can impose regulatory burdens on cable television operators, because of their potential adverse impact on broadcast television. Such ancillary jurisdiction operates from the view that if the FCC can regulate broadcasting, then it can regulate other media that impact its broadcast regulatory mission. The ability of cable television to migrate audiences and revenues from broadcasters served as the justification for FCC regulations designed to limit the harm. For example, despite having First Amendment rights to program their channel capacity without government influence, cable television operators must reserve up to one-third of their capacity for the mandatory carriage of local broadcast television signals.

The cable television regulatory model applies to a technology that does not directly use radio spectrum. It provides an example of how government might regulate a new technology simply because of the potential for economic harm to operators and consumers of an incumbent technology. Like common carriage and broadcast regulation, such a regulatory construct for cable television can empower governments to mandate public access and universal service responsibilities, albeit at the price of lost revenues and speaker freedoms.

Publishers, including distributors of published content, historically have enjoyed the greatest insulation from government intrusion. For example, in the United States no federal, state or local government agency can compel a newspaper to provide space for an individual to respond to an accusation, press account or editorial. The First Amendment favors publishers' freedom of expression over government intrusion lest publishers self-censor, or otherwise curb their expressions for fear of government reprisal.

Of all pre-existing models, except for non-regulation, the print model provides the greatest safeguards for maintaining a climate conducive to freedom of expression, entrepreneurship, robust investment and risk taking. On the other hand, a bias in favor of non-intervention may so encourage a libertarian environment as to facilitate criminal, predatory and harmful practices most likely to be applied to individuals least able to protect themselves.

THE CLINTON ADMINISTRATION MODEL

In 1997 President Clinton and Vice President Gore released a document designed to articulate a strategy for promoting the Internet by balancing the need for government stewardship without interfering with the benefits accrued from largely unfettered entrepreneurship and ease of market access (Clinton and Gore, 1997). The report suggested principles for application by all governments:

(a) The private sector should lead: support market-driven Internet development, rather than extend preexisting regulatory models; where regulation is necessary, support industry self-regulation and private sector leadership before adopting government mandated solutions.

(b) Governments should avoid undue restrictions on electronic commerce. As the Internet promises to operate as a major new medium for commerce, parties to the transaction should be able to enter into binding agreements for the sale of goods and services with minimum government involvement or intervention.

(c) Where government involvement is needed, its aim should be to support and enforce a predictable, minimalist, consistent and simple legal environment for commerce. Governments will need to enforce competition (antitrust) policies, protect intellectual property rights, safeguard privacy interests, prevent fraud, provide forums for dispute resolution and litigation and foster transparent, readily understood operating rules.

(d) Governments should recognize the unique qualities of the Internet. The lack of unilateral and heavy handed regulation coupled with the populist, bottom-up nature of Internet self-governance has created an environment conducive to entrepreneurship and innovation that should continue.

(e) Electronic commerce over the Internet should be facilitated on a global basis. Governments will need to coordinate the development of a universal and understandable legal framework for global electronic commercial transactions.

PROBLEMS OF REGULATORY ASYMMETRY

Any regulatory regime applied exclusively to Internet applications runs the risk of creating a dichotomy in regulatory rights and responsibilities between providers of functionally equivalent services. Many of the services available via the Internet provide a faster, better, cheaper and smarter evolution of pre-existing services. The Internet provides a convenient, user-friendly medium for acquiring news and entertainment and for engaging in all sorts of commercial transactions. A bias or intention not to regulate, or to regulate lightly such activities may contrast significantly with a pre-existing and more intrusive regulatory model. Governments should not automatically extend the application of legacy regulatory regimes to Internet-mediated equivalent services. Nor should governments deregulate incumbent services simply because Internet options have become available and they have opted to apply a different and probably less burdensome regulatory regime to Internet services.

The onset of Internet-mediated services does present a regulatory challenge to governments, particularly those disinclined to treat such services as equivalent to services transmitted and delivered via traditional media. The juxtaposition of different regulatory regimes typically also creates an asymmetry that has the potential for tilting the competitive playing field in favor of the less regulated service. To the extent that regulation can impose financial and operational burdens, the service provider subject to greater regulation typically suffers a competitive disadvantage vis-à-vis the less regulated operator. Governments need compelling justifications for establishing different regulatory regimes in view of the potential for such asymmetry to impact the marketplace attractiveness of one service vis-à-vis others.

Regulatory dichotomies work best when technological categories remain discrete and absolute. But they certainly do not work when technological convergence results in porous service categories and diversification by operators. When cable telephone and ISPs offer telephone services functionally similar to that available from telephone companies, regulators may not be able to maintain pre-existing dichotomies. Heretofore, government regulators have assumed that incumbent telephone service providers have dominant market shares, should operate as common carriers and offer the best technology and wherewithal to achieve universal service goals. Government regulators typically assume that market entrants like ISPs, other enhanced service providers and resellers of basic transmission capacity do not have the potential to acquire a dominant market share, or that they offer ancillary, non-common carrier services. While incumbent telephone companies incur significant financial duties to serve costly remote areas, the newcomers enjoy exemptions from having to pay charges for accessing the public switched telephone network (PSTN) and from contributing to universal service funding. These ventures qualified for such exemptions on the grounds that they did not offer telephone service even though their offerings might require access to the PSTN.

When ISPs offer consumers telephone service equivalents, which link PSTN access with Internet-mediated telephony, pre-existing regulatory exemptions tilt the competitive playing field to their advantage. Should significant telephony traffic volumes migrate to routings exempt from universal service contribution requirements, the sum of funds available to achieve the universal service mission will decline. The potential for declining universal service funds occurs just as many governments have articulated a broader and more ambitious universal service mission for all citizens to have access to both basic telephone service and advanced Internet services.

REGULATORY OPPORTUNISM

Some providers of Internet-mediated services enjoy the opportunity to provide competitive, functional equivalents to regulated offerings without the same regulatory burdens. Without adjustments in the legal and regulatory arena, these ventures, typically market entrants, may achieve commercial success without having developed a faster, better and more efficient and convenient innovation. They may offer something technologically and operationally awkward, but nevertheless cheaper, because regulatory classifications exempt the operator from having to pay regulator-imposed fees.

Legislative changes in telecommunications laws occur most infrequently, and regulatory lag frequently creates a significant time period during which changed technological and marketplace conditions increasingly contrast with the regulatory status quo.[12] During such periods of delayed adjustment the regulatory process may favor one competitor over others. This is most likely to occur when marketplace conditions trigger new competitive opportunities and when technological convergence eliminates barriers to market entry or market segmentation, such as a separate wire for telephone and video service.

UNITED STATES ATTEMPTS TO PROMOTE COMPETITION AND CONVERGENCE

The authors of the Telecommunications Act of 1996 in the United States belatedly acted with an eye toward recalibrating the telecommunications regulatory regime so that more robust competition might ensue. The law sought to remove barriers to market entry, but in a way that forecloses market power abuses and other anticompetitive practices by incumbents. To the apparent dismay of Congress, telecommunication and information service providers have proven themselves quite adept at exploiting opportunities to capture greater profits and market share not as a function of superior business acumen and efficiency, but simply by exploiting regulatory loopholes that create an uneven competitive playing field.[13] Congress had envisioned a more robustly competitive telecommunications marketplace, but made little reference to whether and how such a diverse group of players might fit within one or more regulatory classifications, particularly in light of the fact that open entry would result in competition by operators having different degrees of market share and financial resources. In the absence of Congressional action, the FCC had to devise and apply appropriate but perhaps different regulatory treatment, including the imposition of common carrier regulation on a subsidiary of an enterprise historically exempt from such regulation.

Some telecommunications ventures have avoided costly regulatory burdens simply on grounds that they lack market power, or because they have semantically crafted services so that they qualified for little or no regulatory oversight. On the other hand some incumbents have continued to incur such burdens despite changed circumstances and the requirement contained in the Telecommunications Act of 1996 that all service providers, regardless of regulatory classification, should bear regulator-imposed financial burdens on a equitable and nondiscriminatory basis.[14]

Both newcomers and subsidiaries of incumbents may secure regulatory exemptions on semantic grounds, that is, by characterizing and offering services in a way that qualifies for diminished regulation. Incumbents may exploit regulatory inertia that maintains regulatory safeguards designed for natural monopolies operating in noncompetitive environments, which benefit the incumbent perhaps to the detriment of its consumers and the broad, elastic concept of the public interest, convenience and necessity.[15]

Such regulatory arbitrage often triggers micro-level concerns in the pleadings filed in proceedings at the FCC over, for example, the jurisprudential and marketplace consequences of private and common carriers providing functionally equivalent services subjected to quite different regulation.[16] While appropriate for their purposes the authors of such materials miss opportunities for greater benefits: a relatively narrow FCC rulemaking proceeding provides little opportunity to address and resolve broader concerns about regulatory consistency (see, for example, Krattenmaker and Powe Jr, 1995).

CASE STUDIES IN REGULATORY ASYMMETRY

Internet Telephony

The Internet's ability to switch and route audio packets in real time makes it possible for entrepreneurs to offer the functional equivalent to telephone service. The interconnected and integrated networks that comprise the Internet can stream digital packets constituting a live voice conversation. Packeting voice conversations may constitute an efficient alternative to the conventional circuit-switched systems currently in use. However, the opportunity to evade regulatory policies and fees makes Internet telephony cheaper regardless of whether the technology in actuality achieves greater efficiencies (Frieden, 1997b). Having conscientiously sought to refrain from regulating the Internet, regulators devised classifications that differentiate basic telecommunications from enhanced services, including ones delivered via telecommunications links. Words like enhanced, value-added, information and interactive computer refer to services that use telecommunication as building blocks for transporting a service that presumably does not constitute an alternative to such basic voice services.

A basic versus enhanced services dichotomy breaks down when providers of enhanced services do indeed provide unregulated or lightly regulated functional equivalents to what heavily regulated common carriers offer. In addition to evading the regulatory burdens and reduced operational flexibility, Internet telephony entrepreneurs also avoid having to pay the above cost fees borne by long distance carriers for use of local telephone company 'first and last' mile exchange facilities. Additionally Internet telephony providers bear no financial obligation to contribute to universal service funding, which in some nations now includes subsidies to make the Internet accessible in schools, libraries and medical facilities.

Notwithstanding the adverse financial impact on its universal service mission, the FCC remains adamantly opposed to extending traditional telecommunications regulation to ISPs (Oxman, 1999). To its credit, the Commission has embraced a predisposition not to extend its regulatory wingspan to include Internet-mediated services. The Commission believes market forces will create incentives for a robustly competitive and ubiquitous high-speed information service infrastructure. Likewise, it wants to apply the pre-existing regulatory dichotomy between regulated, common carrier telecommunications and unregulated, non-common carrier enhanced services articulated in the *Computer Inquiries*. Also, the FCC wants to support the Clinton Administration's view that the Internet should be a tax-free, largely unregulated medium.

In a larger sense the FCC, through a staff working paper, has expressed its reluctance to extend the common carrier classification and the regulatory burden it generates on Internet-mediated services, including ones that compete with, and appear as viable alternatives to, common carrier services. The Commission appears disinclined to impose legacy regulations on new technologies even if these technologies migrate traffic and revenues from services that have borne the universal service subsidy obligation:

> New technologies, while perhaps similar in appearance or in functionality, should not be stuffed into what may be ill-fitting regulatory categories in the

name of regulation. Rather, the Commission should continue the approach of studying new technologies and only stepping in where the purpose for which the Commission was created, protecting the public interest, demands it. (Oxman, 1999, pp. 24–5).

The FCC's in-house think tank favors deregulating incumbents rather than regulating market entrants. The FCC should pay attention to the potential for anticompetitive conduct, and adverse impact on universal service funding. However, the Commission should decide to apply regulatory safeguards on an ad hoc, as needed basis and for instances where regulatory intervention outweighs the costs imposed.

Devising New Legal and Regulatory Safeguards

Notwithstanding the predisposition to eschew regulating the Internet, legislatures and regulators cannot avoid the fact that Internet mediation can trigger undesirable outcomes. Without some government intervention, children might have access to content deemed undesirable by most citizens. Without regulatory safeguards, purveyors of harmful or undesirable content might have easy and lawful access to children. On the other hand, ISPs seeking to offer safeguards, including speedy blockage or removal of such content might face civil liability to the content provider. Similarly, without legislative-conferred exemptions, an ISP might incur liability for serving as a conduit for the distribution of content that infringes intellectual property rights, despite the lack of knowledge and the ability to know that such infringement has taken place.

In the absence of specific legislative or regulatory relief, ISPs run the risk of incurring substantial fines and even criminal culpability for serving as a third party in an Internet-mediated transaction. In the United States, Congress exempted ISPs from liability for taking reasonable action to protect children by blocking or screening offensive material. Section 230 of the Communications Act of 1934, as amended, exempts interactive computer service providers from civil liability when serving as 'good Samaritans' and taking steps in good faith 'to restrict access to or availability of material that the provider or user considers to be obscene, lewd, lascivious, filthy, excessively violent, harassing or otherwise objectionable, whether or not such material is constitutionally protected'.[17] Section 230 of the Communications Act provides ISPs with broad immunity from civil liability for serving as a conduit for harmful speech in much the same way as the traditional common carrier regulatory construct; that is, that operators of neutral, transparent conduit providers do not operate as publishers or speakers of the content transmitted via their facilities.[18] Accordingly, this Section exempted an ISP from liability when it first served as a conduit for the dissemination of hateful and defamatory messages and for the manner and timing of its investigation. In *Zeran v. America Online, Inc.* a major on-line service provider incurred no liability for carrying messages that applauded the bombing of a federal government building and attributed such hateful messages to an individual who did not post them.[19] The plaintiff, Zeran, claimed that America Online acted negligently in failing to block messages attacking him and for refusing to issue a retraction. The reviewing federal court held that Section 230 of the Communications Act foreclosed the application of state laws and regulations pertaining to defamation and thereby exempted America Online from any liability for its operation as a conduit for tortuous speech defaming Zeran and for allowing additional attacking messages to be posted during the time it did nothing to correct the mistaken conclusion that Zeran supported the office bombing.

Without legislative or regulatory relief, ISPs face similar potential liability when serving as the conduit for distributing content that violates intellectual property rights. The digital nature of the Internet makes it possible to deliver identical duplicates of valuable content in video, graphical or audio formats. ISPs serve as unknowing conduits for the illegal delivery and distribution of pirated software, music, photographs and books. Without legislative or regulatory exemptions, ISPs could incur direct, contributory or vicarious liability for serving as third party distribution conduits. Arguably some of the basic technological functions of computers and the Internet make copies, in caches and temporary files, which under traditional laws and regulations might constitute infringement.

In the United States, Congress enacted the Digital Millennium Copyright Act (DMCA) to incorporate international initiatives, under the auspices of the World Intellectual Property Organization, to protect valuable intellectual property from new technological pirating vehicles. This legislation and other bills in Congress seek to refine the intellectual property protection in view of new ways to protect against infringement, but also to circumvent these new technological protections. Additionally the DMCA provides on-line operators with exculpation from liability when they accrue no direct financial benefit from copyright violating activity and take affirmative steps to remove or disable such infringing actions.[20] Legislation provides a model for narrowly tailoring a remedy to new problems resulting from Internet mediation. However, one cannot expect frequent legislatively conferred solutions given the expense, logistics, and other issues competing for attention.

THE FALLACY OF AN UNREGULATED INTERNET

Most Internet developers, engineers and designers embrace the view that a significant part of the Internet's success stems from the decision by most governments to eschew regulation. Clearly governments have conscientiously avoided the automatic decision to impose or extend regulations simply on grounds that the Internet provides a medium for some previously regulated activity. Even at the risk of creating a regulatory loophole and an opportunity to avoid otherwise applicable regulatory burdens, governments perceived real benefits in incubating and promoting Internet development.

The Internet has so developed and matured that many governments have withdrawn or tapered financial support. Similarly, some of the previously small regulatory loopholes have grown in importance and potential for financial harm. If the Internet reaches a critical mass and provides a direct competitive alternative to regulated, taxed and government scrutinized services, then it will become increasingly difficult to maintain a hands-off approach. Also we should appreciate that governments never intended tax moratoria and deregulation to provide a vehicle for individuals or businesses to engage lawfully in conduct that otherwise would be considered illegal or harmful.

Internet mediation does not by itself legitimize illegal conduct, or insulate one from civil liability. For example many nations and states deem illegal gambling, procuring medicine without a doctor's prescription, purchase of alcohol by a minor, unreported transfers of financial instruments, or accessing obscenity. Yet these and other activities might not violate the laws and regulations of some other jurisdictions. At least in theory one might claim to have engaged in lawful activity simply by using the Internet as a medium to avail oneself of the liberal laws and regulations in some jurisdictions regarding gambling, alcohol, money laundering, and access to medicine, pornography and other controlled goods and services. However, just because an activity might be legal in one jurisdiction does not, by itself, make it a lawful activity in another jurisdiction. Put another way, one may not evade one nation's laws, policies and regulations simply by 'virtually' traveling via the Internet to another nation.

JURISDICTIONAL ISSUES

Internet pundits glibly proclaim the lack of physical place in cyberspace. Yet legal and business conventions require the identification of physical location for determining which nation has jurisdiction over the activity. In many instances, parties to a contract can specify which national, state or provincial law shall apply, but in other cases the nature of the transaction muddies a clear choice of law. In the physical and pre-Internet world, businesses and individuals could access and apply favorable laws, regulations and policies simply by taking some steps to specify an attractive location as having jurisdiction over the transaction. For example an individual, resident in a state or nation that outlaws gambling, might travel to Monte Carlo or Aruba for that kind of action. In many instances, one need only establish a tentative or even temporary presence to secure access to such favorable conditions. One can establish a bank or trading account in places like Switzerland and thereby secure the application of favorable privacy rules. Simple and less burdensome incorporation laws in places like Bermuda and the Cayman Islands make it possible for companies to avail themselves of favorable commercial and corporate laws with little more than a mailing address.

Internet-mediated transactions confound pre-existing legal precedents primarily by generating confusion over which laws apply and which entity has jurisdiction. Two decidedly different outcomes can occur:

(a) *Remote access/going to a jurisdiction:* A business or individual can go to a jurisdiction remotely, that is, without establishing any physical presence in a jurisdiction, consumers and businesses can have remote access to and apply that nation's or state's laws; or

(b) *Remote application/coming from a jurisdiction:* The laws, regulations and policies of the country or state apply to non-resident businesses or individuals even for Internet-mediated commerce, that is, without establishing any physical presence in a jurisdiction, one can become subject to its national or state laws.

Remote access extends the pre-existing opportunities for individuals and businesses to access laws, regulations and policies more favorable than those the host nation or state otherwise would have applied. In some instances, using the Internet to go to a jurisdiction makes it possible to evade laws applicable to residents and to engage in transactions deemed illegal by the nation or state that otherwise would have had jurisdiction over the transaction. For example, one might use Internet mediation as a vehicle to evade a federal or state prohibition on gambling, to access types of pornography that would have constituted obscenity, to launder money or purchase goods (alcohol, tobacco, firearms, drugs and medications) and services without proper licenses, tax and duty payments, or other types of authorizations.

Since the beginning of international commerce, businesses and individuals have exploited remote access opportunities. For example, ocean going shippers and cruise lines have operated under flags of convenience, having secured licenses to operate, favorable tax treatment and possibly greater opportunities for exculpation from liability for offenses that elsewhere would have triggered civil and criminal liability. The Internet provides a more robust and easier medium for exploitation of such loopholes; for example, on-line incorporation in 20 minutes.

Remote application of a nation's laws threatens to apply extraterritorially the kinds of laws, regulations and policies that businesses and individuals seek to evade.[21] Worse yet, one might not even know that just one single, transaction might trigger the application of jurisdiction and exposure to civil and criminal liability. For example, the operator of a World Wide Web site in the United States might not offer the level of privacy protection legislated in the European Community. In the likely event that a European Web surfer happens on and transacts commerce with this United States-based site, a European court, regulator or other institution might find a violation and impose civil liability including monetary forfeitures on any number of participants in the transaction, such as the ISP and the credit card companies processing the transaction in addition to the seller of the goods or services.

Remote application of laws, regulations and policies poses the possibility of extraterritorial jurisdiction on such diverse areas as privacy and data protection, other consumer protection laws, professional and business licensing, labor laws, intellectual property protection, taxation, and laws applicable to particular types of transactions, such as gambling, adult material, hate speech, and defamation. Efforts already have begun to harmonize laws and establish procedures that promote electronic commerce. Yet such procedures upset some free market advocates, and may prove ineffectual to the small and nimble entrepreneurs keen on exploiting the Internet's market accessibility to engage in gray and black market activities:

[While the] Internet does not pose unique problems for enforcement against elephants [i.e. established, law-abiding players, it] . . . does pose new problems for the legal regulation of mice, who are difficult to find and often hide outside of the country. Because the Internet helps individuals conduct international transactions with unparalleled ease, the harms caused by these far-off mice are potentially large [and tempt] . . . [c]ountries that cannot catch the mice . . . to enact laws regulating other parties, such as individual consumers, ISPs, payment providers, or other countries where the mice hide. (Swire, 1998)

CONCLUSION

Internet mediation by itself provides neither an exemption from, nor a requirement for regulation of online activity. Personal or corporate behavior that triggered regulation, civil liability or criminal culpability without Internet mediation typically would do so where the Internet provides the link. Insofar as the nature and scope of regulation is concerned, one should consider the Internet as simply one of many media for linking individuals and businesses. The nature of the linkage and particular characteristics of the Internet will present novel and challenging regulatory and legal challenges, but these challenges typically favor extension of pre-existing case precedent and regulatory models.

Internet mediation does, however, present instances where citizens and their representative governments do not like the outcome reached by extension of the legal and regulatory status quo. In those instances, the legislature must intercede and erect an ad hoc remedy. For example in the last two years the United States Congress has enacted several laws that address problematic outcomes resulting from the application of existing law including the following.

- Internet Tax Freedom Act: Provides a three-year moratorium on new taxes for Internet access charges. A few states that currently tax Internet access would be permitted to retain those taxes, but no state may enact a new levy; prohibits multiple or discriminatory taxes on electronic commerce, including Web search taxes, e-mail surcharges, taxes on out of state Web sites, and other levies that target the Internet; creates a temporary Advisory Commission on Electronic Commerce to study electronic commerce tax issues and report back to Congress; and directs the President to work in international bodies for the establishment of the Internet as a global tariff-free trade zone.[22]
- Children's Online Privacy Protection Act: Requires companies to obtain parental permission before collecting any personal information from children under the age of 13.[23]
- Child Online Protection Act: Requires operators of websites containing material that is harmful to minors, to verify that website viewers are adults and to restrict access to minors.[24]
- Government Paperwork Elimination Act: Establishes authentication standards for electronic communications (digital signatures) used by the federal government.[25]
- Digital Millennium Copyright Act: Establishes limited liability for on-line copyright infringement for entities offering the transmission,

routing, or providing of connections for digital on-line communications between points specified by a user of material of the user's choosing, without modification of the material; and providers of online services or network access.[26]

The legislation outlined above evidences willingness by government to change the rules when presented by technological innovations that, for example, provide new or more powerful means to promote commerce, harm children, govern and evade intellectual property rights protection. No doubt the list of legislative and regulatory remedies will grow, but governments appear keen on applying analogies and extrapolating from current models wherever possible.

NOTES

1. Oxman (1999) advocates the deregulation of pre-existing telecommunication services when Internet-based services provide competitive alternatives rather than the imposition of legacy regulations on such new technologies.
2. See, e.g., The Western Union Tel. Co. v. Esteve Bros, 256 U.S. 566 (1921) (exculpatory clauses in common carrier tariff limited liability to refunding cost of carriage despite substantial financial damage resulting from non-delivery of an message transmitted only once). For an examination of exculpation of common carrier liability see Kunin (1992).
3. See, e.g., Bell System Tariff Offerings, 46 FCC 2d 413 (1974), *affirmed sub nom.* Bell Telephone Co. of Pa. v. FCC, 503 F.2d 1250 (3d Cir. 1974), *cert. denied*, 422 U.S. 1026 *rehearing denied*, 423 U.S. 886 (1975); MCI Telecommunications Corp. v. FCC, 580 F.2d 590 *cert. denied*, 439 U.S. 980 (1978) (access to local exchange facilities mandated); Establishment of Domestic Communications Satellite Facilities, 22 FCC 2d 86, 97 (1970), *policy reaffirmed*, 34 FCC 2d 9, 64–5, *adopted*, 35 FCC 2d 844, 856 (1972), *on reconsideration*, 38 FCC 2d 665 (1972) (domestic satellite policy mandates non-discriminatory, diverse, and flexible access to domestic satellites and earth station facilities); *accord* Specialized Common Carrier Services, 29 FCC 2d 870, 940 (1971) (AT&T required to afford local exchange facility access to competing inter-city carriers), *on reconsideration*, 35 FCC 2d 1106 (1971), *affirmed sub nom.*, Washington Utilities and Transportation Comm. v. FCC, 513 F.2d 1192 (9th Cir.), *cert. denied* 423 U.S. 836 (1975). Use of the Carterfone Device in Message Toll Tel. Serv., 13 FCC2d 420, *recon den.*, 14 FCC2d 571 (1968)(invalidating local exchange carrier tariff restrictions on interconnection of customer premises equipment with the telephone network); Telerent Leasing Corp., 45 FCC 2d 204 (1974), *aff'd sub nom.* North Carolina Utilities Commission v. FCC, 537 F.2d 787 (4th Cir.), *cert. Den.*, 429 U.S. 1027 (1976); Terminal Equipment Registration, 56 FCC 2d 593 (1975), 58 FCC 2d 736 (1976), *aff'd sub nom.*, North Carolina Utilities Commission v. FCC, 552 F.2d 1036 (4th Cir.) *cert. Den.*, 434 U.S. 874 (1977) (pre-empting the states on the matter of customer premises equipment interconnection with the telephone network).
4. 'A firm controlling bottleneck facilities has the ability to impede access of its competitors to those facilities. We must be in a position to contend with this type of potential abuse. We treat control of bottleneck facilities as prima facie evidence of market power requiring detailed regulatory scrutiny. Control of bottleneck facilities is present when a firm or group of firms has sufficient command over some essential commodity or facility in its industry or trade to be able to impede new entrants. Thus bottleneck control

describes the structural characteristic of a market that new entrants must either be allowed to share the bottleneck facility or fail'. Policy and Rules Concerning Rates for Competitive Common Carrier Services and Facilities Authorizations Therefore, CC Docket No. 79–252, First Report and Order, 85 FCC 2d at 36. See, also, United States v. Terminal Railroad Association, 224 U.S. 383 (1912) (antitrust court ordered railroads to provide competitors equivalent access to bottleneck railway terminal facilities), *appeal after remand*, 236 U.S. 194 (1915); Cellular Communications Systems, 86 FCC2d 469, 495–96 (1981) (Commission required telephone companies to furnish interconnection to cellular systems upon terms no less favorable than those used by or offered to wire-line carriers), *modified*, 89 FCC2d 58 (1982), *further modified*, 90 FCC2d 571 (1982); Need to Promote Competition and Efficient Use of Spectrum for Radio Common Carrier Services, 59 RR2d 1275 (1986), *clarified*, 2 FCC2d 2910 (1987), *aff'd on_recon.*, 4 FCCRcd 2369 (1989) (Commission clarified policies regarding interconnection of cellular and other radio common carrier facilities to landline network); Lincoln Tel. & Tel. Co., 659 F.2d at 1103-06 (court upheld Commission's order requiring Lincoln to provide interconnection facilities to MCI); MCI Telecommunications Corp. v. FCC, 580 F.2d 590 (D.C.Cir.), *cert. denied*, 439 U.S. 980 (1978); Bell Tel. Co. of Pennsylvania v. FCC, 503 F.2d 1250(3d Cir. 1974), *cert. den.* 422 U.S. 1026 (1975) *reh. den.,* 423 U.S. 886 (1975).

5. In Sable Communications, Inc. v. FCC, 492 U.S. 115 (1989) the Supreme Court upheld a federal statute prohibiting obscene telephone messages, but overturned the statute's absolute denial of adult access via telecommunication common carriers to indecent messages that are entitled to First Amendment protection.

6. In FCC v. Midwest Video Corp., 440 U.S. 689 (1979) the United States Supreme Court struck down FCC rules requiring cable television operators to set aside channel capacity for public, educational and government use on grounds that cable television does not constitute common carriage.

7. See Turner Broadcasting Sys, Inc. v. FCC, 114 S.Ct. 2445 (1994).

8. Telecommunications Act of 1996, 104 P.L. 104, 110 Stat. 56, *codified at* 47 U.S.C. Sec. 151 *et seq.*

9. Sec. 153(44) defines telecommunications carrier 'as a common carrier . . . to the extent that it is engaged in providing telecommunications services'. Sec. 332(c)(1)(A) also requires the FCC to treat as common carriage the provision of commercial mobile services.

10. Section 160(a) of the revised Communications Act orders the FCC to 'forbear from applying any regulation or any provision of [the Communications Act] to a telecommunications carrier or telecommunication service' if such regulation is no longer necessary to ensure just, reasonable and nondiscriminatory rates, to safeguard consumers and that such forbearance would serve the public interest.

11. Historically, common carriers have operated as neutral and transparent conduits, neither knowledgeable of the content they carrier, nor legally responsible for what they carry. The Telecommunications Act of 1996 also provides legal protection for the 'Good Samaritan' blocking and screening of offensive material defined as 'any action voluntarily taken in good faith to restrict access to or availability of material that the provider or user considers to be obscene, lewd, lascivious, filthy, excessively violent, harassing, or otherwise objectionable, whether or not such material is constitutionally protected'. Protection for 'Good Samaritan' Blocking and Screening of Offensive Material, 47 U.S.C. sec. 230(c).

12. Regulatory lag has been defined as the general delay in the responses of regulators to changes in cost or market conditions (Crandall and Sidak, 1995).

13. The following uses the term in the context of telecommunication regulation: 'As the CMRS market becomes more competitive, incumbent LECs may have an incentive to tilt the CMRS playing field in their favor. Our response in this Report and Order is to establish several safeguards that minimize the opportunity for incumbent LECs to recover their CMRS costs from local telephone customers. Local telephone customers should pay for local telephone service, not for CMRS service'. Separate Statement of Commissioner James Quello in Amendment of the Commission's Rules to Establish Competitive Service

Safeguards for Local Exchange Carrier Provision of Commercial Mobile Radio Service, WT Docket No. 96-162, 1997 West Law 609265 (FCC) (3 October 1997). See also Frieden (1997a).

14. For example, the 1996 Act requires, '[a]ll providers of telecommunications services . . . [to] make an equitable and nondiscriminatory contribution to the preservation and advancement of universal service', 47 U.S.C. sec. 254(b)(4).

15. For example, in regulating broadcasters to prevent interference, the FCC may order changes in frequencies, power and operating times on determining 'that such changes will promote public convenience or interest or will serve public necessity' 47 U.S.C. sec. 303(f).

16. Regulatory arbitrage involves the ability to exploit differences in regulatory treatment to one's financial and competitive advantage by gravitating to jurisdictions or regulatory classifications conferring favorable treatment. See Froomkin (1997).

17. Protection for Private Blocking and Screening of Offensive Material, Communications Act of 1934, as amended, Section 230(c)(2)(A), *codified at* 47 U.S.C. sec. sec. 230(c)(2) (A)(1999).

18. Section 230 makes this point explicit: 'No provider or user of an interactive computer service shall be treated as a publisher or speaker of any information provided by another information content provider'. 47 U.S.C. sec. 230(c)(1).

19. 958 F.Supp. 1124 (E.D. Va. 1997), *aff'd*, 129 F.3d 327 (4th Cir. 1997)., *cert. den.*, 118 S.Ct. 2341(1998).

20. Digital Millenium Copyright Act, *codified at* 17 U.S.C. sec. 512 (1999) and elsewhere.

21. See Swire (1998), Goldsmith (1998) and Stein (1998).

22. Pub.L. 105-277, 112 Stat. 2681-719, *codified at various sections* (1999).

23. Pub.L. 105-277, 112 Stat. 2681-728, *codified at* 15 U.S.C. sec. 6501 *et seq* (1999).

24. Pub.L. 105-277, 112 Stat. 2681-736, *codified at* 47 U.S.C.A. sec. 231 (1999).

25. Pub.L. 105-277, 112 Stat. 2681-749, *codified at* 44 U.S.C. sec. 3501 *et seq* (1999).

26. Pub.L. 105-304, 112 Stat. 2860, *codified at various sections* (1999).

REFERENCES

Clinton, W.J. and Gore, A. Jr (1997), 'A Framework for Global Electronic Commerce', available at http://www.ecommerce.gov/framewrk.htm.

Crandall, R.W. and Sidak, J.G. (1995), 'Competition and Regulatory Policies for Interactive Broadband Networks', *Southern California Law Review*, 68, 1203–20.

Frieden, R.M. (1995), 'Contamination of the Common Carrier Concept in Telecommunications', *Telecommunications Policy*, 19, 685–97.

Frieden, R.M. (1997a), 'The Telecommunications Act of 1996: Predicting the Winners and Losers', *Hastings Communications and Entertainment Law Journal*, 20, 11–57.

Frieden, R.M. (1997b), 'Dialing for Dollars: Will the FCC Regulate Internet Telephony?', *Rutgers Computer and Technology Law Journal*, 23, 47–79.

Froomkin, A.M. (1997), 'The Internet as a Source of Regulatory Arbitrage', in B. Kahin and C. Nesson (eds), *Borders in Cyberspace: Information Policy and the Global Information Infrastructure*, Cambridge, MA: MIT press, pp. 129–47.

Goldsmith, J. (1998), 'What Internet Gambling Legislation Teaches about Internet Regulation', *International Lawyer*, 32, 1115.

Krattenmaker, T.G. and Powe, L.A. Jr (1995), 'Converging First Amendment Principles for Converging Communications Media', *Yale Law Journal*, 104, 1719.

Kunin, C.C. (1992), 'Unilateral Tariff Exculpation in the Era of Competitive Telecommunications', *Catholic University Law Review*, 41, 907.

Oxman, J. (1999), 'The FCC and the Unregulation of the Internet', Federal Communications Commission, Office of Plans and Policy, Working Paper No. 31, available at http://www.fcc.gov/opp/workingp.html.

Stein, A.R. (1998), 'The Unexceptional Problem of Jurisdiction in Cyberspace', *International Lawyer*, 32, 1167.

Swire, P.P. (1998), 'Of Elephants, Mice and Privacy: International Choice of Law and the Internet', *International Lawyer*, 32, 1991.

7. Product bundling and wholesale pricing

Timothy J. Tardiff

INTRODUCTION

The demand for telecommunications services is growing along many dimensions. Not only are the volumes of old services growing at healthy rates, for example, voice traffic increases at a rate of 6 per cent to 7 per cent annually (Lenahan, 1999), new services have emerged and are growing even more rapidly. For example, in the United States (US), volumes of data traffic now exceed voice, and these volumes are growing at an annual rate of 45 per cent.[1] As a result, consumers must select from a proliferating array of old and new services, as well as different payment options. Further, firms that formerly operated in separate markets are competing by offering packages of services that cut across the increasingly indistinct boundaries between these markets. Examples include the packaging of local and long-distance traffic; offering of both voice and high-speed data services, sometimes over the same telephone line or cable television connection; packaging of wireless and wireline services; and packaging of telephone, video and Internet access over the same high-speed facility.

Clearly these trends are producing complex product offerings and pricing structures. As a result, the familiar prescription (or description) in economics that price equals cost in competitive markets is no longer adequate to describe the richness of competitive outcomes in telecommunications. In order to understand these outcomes and the concomitant pricing patterns, it is necessary to consider the cost structures of firms offering these services. Cost structures and their pricing implications are discussed in the next section. Based on this framework, examples of pricing trends for retail telecommunications services are described. These include the growth of pricing options in place of single undifferentiated prices and the offering of 'one-stop shopping' opportunities through the convergence of formerly separate markets. The discussion of retail pricing options is followed by an assessment of whether certain types of pricing responses, particularly on the part of incumbent providers facing entry by new firms and/or incumbent firms from other markets, are pro- or anti-competitive.

COST STRUCTURES AND THEIR IMPLICATIONS FOR PRICING

Cost Structures

Firms that produce telecommunications services differ substantially from those perfectly competitive firms found populating economics textbooks. Rather than employing a known technology to produce an undifferentiated service at constant marginal cost, but with no fixed costs, telecommunications firms (as well as firms in the information economy) generally produce several services and employ technology with distinct cost characteristics (Shapiro and Varian, 1999). Typically, the cost of producing service entails high fixed costs, which are often sunk on investment, but a low unit cost for additional production. For example, a switch used to connect telephone customers within a particular area and customers in other areas imposes a large up-front cost for the basic switch capability, but smaller additional costs as more subscribers in an area hook up to the switch. That much telecommunications investment is sunk is illustrated by the cost of providing new wireline facilities to customer locations. For example, once a trench is dug to carry wires between a switch and customer locations, these costs are often sunk because that trench might not be reused, for example to serve customers in other areas.

High fixed costs and low unit costs, by themselves, imply that telecommunications firms have economies of scale – the average cost per unit declines as volume increases. In addition, such firms have economies of shared production or scope. That is, it is often less expensive for a single firm to produce more than one service than to set up separate firms to produce each service separately. For example, combining separate offerings into a single package allows a firm to experience savings from consolidating formerly separate bills. On the facilities side of the cost picture, a cable television firm can now offer television, telephone and high-speed Internet access to subscribers through a single connection to the residence.

Another characteristic of telecommunications technologies is that although they can have long services lives, such as ten or more years, for some components, particularly those that are electronic or computer based, the costs for these components tend to decline substantially over time. Hausman (1999) reports that the cost of switching has declined by 8 per cent per annum. This characteristic has implications not only for the general trend of prices in telecommunications (obviously downward), but for the pricing strategies firms can employ and how consumers may respond to the product offerings such advances in technology offer.

Pricing Implications

Perhaps the fundamental economic principle guiding markets still subject to regulation (as is certainly the case for telecommunications in most countries) is that prices established by regulation should approximate those that competition would produce if regulation were not needed (Kahn, 1988, Vol. I: 17). These ideal prices themselves are typically thought to be cost-based prices, where the costs in question are the additional cost required to produce the volume of service in question. With high fixed costs in general, and sunk costs in particular, such incremental cost pricing does not allow the firm to recover its total outlays. The incremental cost will be at or close to the unit costs associated with the volume of service in question, while the total cost of the firm also includes the large amount of fixed costs. While the inadequacy of bare incremental cost has gained greater recognition in telecommunications, total service versions of incremental cost have been the simple solution chosen by a growing number of regulators throughout the world – at least in the case of wholesale prices, such as interconnection and unbundled element pricing.[2] These approaches often involve assigning uniform percentages of the fixed costs to incremental costs.[3]

The accuracy of total service approaches in approximating market outcomes depends on the extent to which the entire production volume would be sold in competitive markets at a single price. In fact, at least for retail services, the trend for unregulated telecommunications services has been price differentiation and the proliferation of multiple offerings serving similar functions, rather than undifferentiated services. For example, for long-haul long distance services in the US, not only have regulators ruled that these markets are sufficiently competitive to minimize regulation, but competitors have responded with a wide range of pricing options for essentially the same service – a telephone call at a particular time.[4] Of course, such product and price differentiation is not unique to telecommunications. Shapiro and Varian (1999) describe how firms offering information products such as software and data products tailor their offerings and pricing plans to fit differentiated consumer preferences.[5]

The tendency for technological progress to lower costs for certain components in a telecommunications cost structure is reflected by declining prices over time. Prices in telecommunications have tended to increase less rapidly than output prices in general, and even more so for those services most subject to technological progress and/or increases in competition.[6] To the extent that competition accelerates such trends, it is possible that long-term contracts will replace traditional tariff prices in some circumstances. A major regulatory pricing issue that has accompanied increased local competition is how to price wholesale services such as unbundled sub-scriber lines in the light of tension between the long life of the facilities needed to provide lines and the objective of competitors to obtain their own facilities rather than leased facilities. A typical tariff price, which most cost models approximate in this respect, generates monthly prices that are equivalent to the payments that would prevail under a long-term contract for the facilities (Tardiff, 1999). However, competitors are usually not required to make the commitment a long-term contact would require and technological progress is likely to make such spot prices unsustainable over the long haul. The implications are that tariff-like prices should be part of a long-term contract or that month-to-month prices should anticipate the uncertainties arising from technological and competitive developments. In particular, such prices would be higher in the near term and then decline over the life of the asset in question (Hausman, 1997, 1999; Mandy, 2000).[7]

TRENDS IN TELECOMMUNICATIONS PRICING

Requirements for Profitable Pricing

The characteristics of telecommunications costs as well as the acceleration of competition have fundamentally changed the nature of telecommunication pricing. The evolution of competition, both in the US and elsewhere, is marked by the convergence of industries and the erosion of formerly pro-tected markets. The new freedom, indeed the necessity, to consider entry into new markets carries with it the burden of selecting which markets, ser-vices, and customers will be most profitable to serve – both in terms of pro-tecting existing lines of business and successfully entering new lines.

Successful answers to such questions require a much different approach to service offerings than was necessary in a monopoly environment where homogeneous, ubiquitous services are the rule. Competitive telecommuni-cations markets require much greater differentiation of services and target-ing of services to particular customer types. The questions that need to be answered are: who will use the new product or service; how much of the new service will be used at different price levels; do potential buyers have preferences for particular price structures, for example higher versus lower fixed and usage costs; and do service features other than price matter?

The offering of traditional services typically required answers to only the second of these questions. In contrast, successful business decisions regarding new offerings, which will be made with increasing frequency in a more competitive and less regulated environment, will typically address all these questions. Timely answers are essential, not primarily because regulators (and intervenors) demand high quality, defensible forecasts, but because competitive success depends on good answers. Previous studies of telecommunications demand (see for example, Tardiff, 1998a), as well as the entry strategies that are being used as markets converge have identified several major themes that are important in determining the success of particular offerings and/or competitors. These issues include the importance of price structure; the importance of brand identity; the quality dimensions of product offerings; the need to offer comprehensive packages (one-stop shopping); and the significance of wholesale and retail product offerings.

(i)   Price structure matters

There are many options for pricing telecommunications services or service packages. Services such as wireless could include one-time up front charges for subscriber equipment; fixed monthly charges; per-minute charges for service use. Econometric studies of customer demand, and the decisions of providers and customers have revealed several interesting findings concerning price structure issues. First, consumers have a much higher aversion to one-time charges than a simple amoritization of such charges would indicate. This has been revealed both for basic telephone service[8] (Perl, 1984 and Hausman et al., 1993) as well as wireless services.[9] Second, there is a strong preference for flat-rate pricing over those that also contain usage charges (Train, 1994). This has long been known in the case of residential service, where observers have long lamented the 'irrational' selection of flat rate over measured-rate service by customers who are apparently better off with the latter. Such flat-rate plans appear to be emerging for other services such as dial-up and high-capacity always-on Internet access, and long-distance plans that offer substantial 'free' minutes.

(ii)   Brand identity

A common question telephone company product managers pose in considering competitive responses, regardless of the service in question, is whether particular providers have an advantage, everything else being equal. For example, in the US many customers of local exchange companies still believe that they obtain their service from AT&T, which has a strong positive brand image. Similarly, there is concern that a firm with a 'high-tech' image, such as Microsoft, might start with a positive image (at least before recent antitrust events) should it choose to enter telecommunications markets. Conversely, some firms such as cable television providers were viewed unfavorably as providers of telecommunications services.

Accordingly, the alliance of major cable television providers with AT&T in the US could be interpreted, in part, as a move on the part of the cable television providers to benefit from AT&T's more favorable brand identity in telecommunications markets. The importance of brand preference appears to cut across cultures. In Japan the incumbent provider, NTT, was found to command a price advantage of up to 10 per cent over long-distance entrants (Tardiff, 1995).

(iii) Quality attributes

The essence of product differentiation is varying prices (and price structures) and the features of competitive offerings to appeal to different customer groups. A study of Japanese consumers found the speed of establishing service and whether or not extra digits were needed when dialing a long-distance call were important in establishing the value of long-distance service and the market shares of firms of varying quality on these dimensions (Tardiff, 1995). Similar studies of wireless services identified the value of wireless versus fixed service (the capability of being untethered) and the value of being able to use a telephone while moving at various speeds. Identifying and quantifying these quality dimensions are crucial in ensuring the success of competitive offerings. Although there appear to be many different ways of using the personal communications services (PCS) spectrum, 'high end' uses that approximate current cellular services appear to have much greater appeal than 'low end' uses such as the unsuccessful Telepoint service introduced in the UK.

(iv) One-stop shopping

Competition and market convergence also raise the issue of whether there is a competitive advantage to being able to offer service packages. One of the most contentious competitive issues that the 1996 Telecommunications Act attempts to resolve is whether the prohibition against incumbent local exchange carriers (LEC) offering interLATA tolls places them at a distinct disadvantage when entrants such as AT&T and MCI/WorldCom, are able to bundle both intra- and interLATA calling. Similarly, when telephone companies compete with cable television companies, whether or not customers prefer to buy a package of telephony and video services is important to establish the competitive positions of participants.

Previous studies show that one-stop shopping can be an important competitive advantage. The trend in telecommunications markets is for providers to offer a range of services in an integrated fashion. The provision of long-distance service in the US is an important exception to this trend. To date, the Federal Communications Commission (FCC) has determined that only two incumbent LECs (Verizon in New York and SBC in Texas) have satisfied the requirements of the 1996 Act and therefore are eligible to offer long distance service. While there may be (or has been) a legitimate public policy rationale, such restrictions on bundled offerings can substantially disadvantage the firms subject to them and deny customers substantial benefits. For example, *Telecommunications Reports* (1996) reports that almost 80 per cent of US households would buy bundled services from a single provider. A Bell South study by Smith (1994) indicates that the ability to combine intra- and interLATA toll calls gave interexchange carriers (IXCs) an advantage that is worth a substantial proportion of price. This preference for one-stop shopping cuts across cultures. In a study of Japanese consumers the ability to obtain calling services from a single provider was estimated to be worth 14 per cent of average price (Tardiff, 1995)

(v) Retail or wholesale

The offering of a new service carries with it the possibility (and perhaps even the requirement) that it be offered on both a retail and wholesale basis. The need to address both sales channels arises from legal requirements (both the 1996 Telecommunications Act and regulatory rules seem to require extensive unbundling and resale) and the possibility that wholesale offerings might function more as an additional sales channel than as a competitive threat to retail services. For example, technology can provide alternative versions of call waiting, for example between local and long-distance incoming calls.[10] The questions LECs need to address in evaluating the market potential of such services include whether this service would appeal to a new group of customers and/or divert custom from traditional services; and whether making the service available to other companies on a wholesale basis would be a more effective sales channel as opposed to competing services to the LEC's new and existing offerings. For example, a new call waiting option might appeal to customers who value the distinction between local and long-distance incoming calls, a group not completely served by the current offering. Similarly, the strong brand identity of firms likely to purchase the wholesale service (the IXCs), while eroding the LEC's share of the market, may have a greater potential to expand the market. Therefore, the extra profits (albeit at a reduced level) from wholesale purchases would more than offset profits lost from retail sale reductions.

The basic issue of market expansion versus loss of market share arises whenever the wholesale/retail distribution question is asked. The answer varies on a case-by-case basis. When there is a high subscription to a particular service, such as basic residential access, the market share loss will most likely outweigh the gain from market expansion. There might even be different perspectives within an organization, with wholesale product management seeking an expanding market and retail product management seeing a market share loss. The basic point, however, is that a successful response to competition will be aided by identifying new customers and services that expand the market.

Rationale for One-stop Shopping

The trend towards one-stop shopping is the result of the economies of integration or scope that prevail in telecommunications.[11] They arise whenever facilities, personnel or capabilities can be potentially shared when separate products are offered. Competitive advantage arising from economies of scope is precisely the kind of efficiency advantages that are expected to prevail under competition. Integration is fundamentally a competitive phenomenon, and the efficiency advantages it confers on the integrated firms are socially beneficent. The fundamental competitive principle of freedom of entry means, under conditions of real-world competition, that existing firms are able to integrate into other operations or markets that they can serve.[12] Competition by integration of existing firms into related markets is most likely to be socially productive because it represents an attempt to capture economies of scope, the manifestation of which is the ability of a firm to supply several services at lower costs than if they were to supply them separately. The source of such economies is the possibility – indeed, the pervasive phenomenon – of existing firms having special capabilities – of their physical plant, their managerial or labor forces, technological or marketing skills or reputations – of taking on the provision of additional products or services at incremental costs lower than the costs of setting up systems to supply those additional services separately.[13]

The benefits of vertical integration and one-stop shopping are by no means confined to driving and holding prices closer to present incremental costs. Competition is a dynamic process. It exerts powerful pressures on suppliers to improve productivity and to be innovative. These potential economies arise on both the demand and the supply sides. The former spring from the attractiveness to consumers of one-stop shopping – purchasing expanding bundles of services, at attractive prices, from single, familiar suppliers. On the supply side, there are ubiquitous promised economies of scale and scope. The greater the capacity of switches and transport facilities, the lower unit costs. This means the incremental costs of adding capacity are lower than average costs. Similarly, the use of common facilities permits the offering of additional services at an incremental cost much lower than where provided on a stand-alone basis. Entry into new lines of business at rates above those low incremental costs provides the opportunity to earn a contribution toward common and fixed costs and profits. These economies have dynamic and static aspects. Complementary goods and services are more plentiful and of a higher quality as the number of users of any one of them, such as the basic telephone service, increases. Since consumers prefer a supplier of communications services that gives them access to the largest number of complementary services – Internet access, information services, database access and video on demand – there is an incentive for industry participants, once freed from legal and regulatory barriers, to compete in developing bundles of services.

The full benefits from competition that enhance one-stop shopping cannot precisely be estimated *a priori*. The essential superiority of competition over comprehensive planning results from the pressure it imposes on market participants to improve their efficiency in ways that cannot be predicted with assurance, hence their usual characterization as X-efficiency, and to innovate, with consequences that are by definition unpredictable. However, estimates suggest that the present asymmetrical restrictions on the incentives of RBOCs to offer new services have cost society billions of US dollars (USD) annually in lost consumer benefits (Hausman and Tardiff, 1995a), deny consumers the substantial benefits of one-stop shopping, subject to considerable disadvantage competitors prevented from offering it, and sacrifice economies of scope, thereby artificially inflating production costs.

Recent Examples of Vertical Integration and One-stop Shopping

In fact, the industry has reacted to the recent dramatic technological and regulatory changes with a kaleidoscope of new ventures, typically involving entry into other markets, pre-existing and new – sometimes by companies operating alone, at other times through partnerships or acquisitions – as each attempts to take advantage of the perceived opportunities.[14] All give rise to the prospect of competition. The major US long-distance providers have made large commitments of capital as well – preponderantly in wireless and in service to business customers in concentrated metropolitan areas. Even before the passage of the 1996 Telecommunications Act, AT&T's purchase of McCaw Cellular had allowed it access to local networks covering about half the US, and it had strengthened its position in these areas by winning licenses in 21 major markets in the ensuing PCS auctions, with bids totaling approximately USD1.7 billion.[15]

Apart from wireless service, the major IXCs have ambitious plans for large-scale facilities-based entry. AT&T acquired Teleport, one of the largest operating competitive local exchange carriers (CLECs), and then, even more dramatically, announced its USD48 billion merger with TCI (the then largest provider of cable television service), which was completed in February 1999 (AT&T and TCI, 1998).[16] Barely two months had passed before AT&T took the further step of acquiring MediaOne (completed by mid-year 2000), the fourth largest cable television provider. In terms of local wire services, AT&T stated several years ago that it had already installed more than 100 local switches and special computers for routing traffic (*Wall Street Journal*, 1995),[17] to this would now be added Teleport's 50 (AT&T, 1998) and TCI's huge network of access lines:

'Completion of this merger accelerates our entry into the $21 billion business local service market because we're reducing our dependence on the Bell Companies for direct connections to businesses,' said AT&T Chairman C. Michael Armstrong. . . . 'We're giving customers simplicity, convenience and choice. It's one-stop shopping for local and long-distance service, just for starters,' he said, and

'TCG has more fiber route miles and serves more businesses in more cities than any other competitive local service company,' Armstrong said. 'The strategic value of this merger . . . positions AT&T for growth and undisputed leadership in three of the fastest growing segments of the communications services industry–consumer, business and wholesale networking services', and

TCG, with more than 10,000 miles of fiber optic cable and 50 local switches, is the nation's premier provider of competitive communications services. Its network encompasses more than 300 communities coast to coast. Armstrong said that AT&T also pledges to devote substantial resources to continue the building of facilities in critical markets. (AT&T, 1998)

Similarly, MCI and Sprint have made major commitments to entering the local market bypassing LEC facilities – with both wireless and extensive terrestrial facilities – in order to serve concentrated business markets.[18] MCI entered into a mammoth, USD37 billion merger with WorldCom, which substantially expanded its local exchange presence, because of WorldCom's previous acquisition of the largest operating CLEC, MFS. Just as in the case of AT&T, announcement of that merger was accompanied by confident proclamations of the way in which it would strengthen the ability of the partnership to provide local exchange service with its own facilities (Schiesel, 1998):

Part of the rationale for WorldCom's acquiring MCI was that the combined company could meld its networks to create a seamless system for global communications. The largest expense for MCI, as a long-distance carrier, had been fees paid to local phone companies for beginning and ending calls, and

MCI WorldCom now wants essentially to eliminate those fees for business customers who use the company for local and long-distance calling. For a conversation or data message that travels exclusively on MCI WorldCom's network, rates could decrease by as much as 35 percent.

Like Pacific Telesis (now joined with Southwestern Bell) previously and US WEST (now merged with Qwest) subsequently, Sprint opted for the newer generation PCS as its main wireless platform, as its aggressive recent advertising campaign attests. These wireless assets are among the major reasons given for the attractiveness of Sprint to MCI WorldCom. The proposed (but subsequently dropped) merger of the companies would have, obviously, enabled both to bolster their respective wireless-, cable- and fiberoptic-based offerings at local, interstate and international levels. In justifying this proposed merger, MCI WorldCom's CEO observed,

> The MCI WorldCom–Sprint merger reflects how the world of telecommunications is changing. Demand is shifting from voice to data, from narrowband to broadband, and from wireline to wireless. In this new world, the notion of separate long-distance and local carriers is falling by the wayside. What is evolving is a landscape where robust global telecom companies offer packages of services that include all distance voice, wireless, and high-speed Internet access. (Ebbers, 1999)

The new facilities-based CLECs are by no means limited to the three major long-distance companies. As of March 1999, over 150 CLECs had installed 724 switches throughout the US – the corresponding figures were 139 at the end of 1996 and 328 at the end of 1997 (Engebreston, 1997; New Paradigm Resources Group, 1998).[19] In addition, there are at least 31 ventures by private electric utilities into telecommunications, making use of their rights-of-way, excess fiber capacity and large capital reserves, which make the telephone and/or cable markets appealing to them.[20] By the end of 1999, these competitive local exchange carriers were serving about 5 per cent of the lines and capturing 7 per cent of local service revenues, with larger business customers generally the prime targets (FCC, 2000a). The availability of investment capital to CLECs has been unequivocally demonstrated. In the three years following the passage of the 1996 Telecommunications Act, they have raised USD30 billion of outside capital (Council of Economic Advisors, 1999). In comparison, recent data reported to the FCC show total annual investment by the incumbent local exchange carriers (ILECs) of about USD19 billion.[21] The former figure, covering a three-year period, was over 12 times the amount of capital those companies had raised in the four years before passage of the Act (Gold, 1998; Council of Economic Advisors, 1999).[22]

In the growing area of high-speed access to the Internet, there is even more vigorous activity, both by vertically integrated incumbents in other markets and by entrants. In the US, incumbent local service providers are markedly behind the cable companies in the provision of high-capacity access and are not expected to catch up until at least 2003.[23] Foremost among these cable companies is AT&T, which is also among the largest providers of long-distance and wireless services. Clearly, high-speed access services are a necessary component of service offerings. Even when the focus is narrowed to providers of DSL service the incumbent RBOCs are not the only competitor. Data CLECs served 100000 DSL lines at the end of 1999, or 20 per cent of the total (ALTS, 2000). Thus firms that hardly existed before 1997 (data CLECs provided only 1500 line in 1997) have made greater inroads into the business of the incumbents than the CLECs offering voice service.[24] Further, data CLECs are expected to increase their DSL lines five-fold by the end of 2000 to 500000 lines, which is equal to the combined number of current DSL lines of CLECs and ILECs.

PRICING WHOLESALE SERVICES

Properties of Efficient Wholesale Prices

Efficient prices are based on costs incurred by a firm supplying the elements or the services in question. Prices based on those costs promote efficient consumption of resources (allocative efficiency), provide for efficient competition among alternative suppliers of local exchange service (productive efficiency), and provide the proper signals for efficient entry and introduction of services (dynamic efficiency).[25] Such prices recognize the uncertainty inherent in demand and technological progress that permeate telecommunications markets. This means that prices based on depreciation rates and cost-of-capital values consistent with the certainty of a long-term contract should only be available to competitors willing to commit to such contracts.[26] Alternatively, competitors that insist on leasing inputs for shorter terms should be required to pay prices that reflect the risks faced by the provider of unbundled elements. In other words, shielding entrants from competitive risks does not provide for efficient competition. Rather, it tilts the option to make or buy network components provided to new entrants inefficiently towards the buy choice. Finally, the provision of local exchange facilities and services involves costs that are shared among multiple elements and/or services; that is, like most firms in the industry LECs have economies of scope.[27] And like these competitive firms, the recovery of such shared and common costs should be determined by market conditions and not by some arbitrary, however fair, allocation.[28]

Establishing Efficient Wholesale Prices

(i) Mandatory versus voluntary offerings

The FCC's interpretation of the 1996 Telecommunications Act clearly mandates network unbundling beyond that which would be required by a traditional 'essential facilities' rationale (FCC, 1996). By making such wide-scale unbundling available, the FCC intended to 'jump start' competition by making attractive make-buy options available to entrants. On 25 January 1999, the Supreme Court overturned the FCC's rule stipulating which components of the ILECs networks must be provided to competitors at regulated prices. The FCC's rule had required that the following network elements be made available by all ILECs, without regard to differences in the markets they serve: the local loop (line between a customer and the ILEC's central office); the network interface device; switching; interoffice transport; signaling; operations support systems; and operator services and directory assistance. The grounds for the rejection were that the FCC had not provided a rationale for unbundling consistent with the requirements under the Act that such elements be necessary for competition and that lack of access would impair the ability of entrants to compete. In place of a rationale consistent with the 'necessary and impair' requirements, the FCC instead chose a convenience basis – the mere fact that a CLEC requested an element demonstrates that it is more convenient to obtain that facility from the ILEC and, therefore, its availability would be procompetitive. The FCC's response to the Supreme Court's remand order was essentially to restate its convenience standard (FCC, 1999).

In unregulated markets, dynamic efficiency considerations create a presumption against mandatory sharing of facilities. The exception occurs when such facilities are necessary for competition and a firm owning such facilities has been engaging in anticompetitive conduct – the essential facilities doctrine. When mandatory access is necessary for competition, the essential facilities that must be made available are those that are necessary for firms to compete in retail markets; are controlled by the monopoly provider; and cannot be economically or technically duplicated. In the telecommunications industry, with its history of regulated monopoly of the local exchange, the case for mandatory access is arguably stronger. By contrast with unregulated markets, where mandatory access can dampen incentives to innovate on a going-forward basis, the ILEC facilities in question are less likely to be the product of such innovation. Therefore there is a legitimate trade-off between speeding up competition by mandatory availability of unbundled elements and the impact of mandatory access on the incentives of both incumbents and entrants to invest in the future. With regard to the latter, a compelling economic case can be made for the proposition that the existence of competitors that do not rely on particular ILEC facilities on a market-by-market basis demonstrates that mandatory access is not necessary for competition and, therefore, mandatory unbundling should not be ordered. To arrive at this decision, it is necessary to determine how competitors provide service in specific markets.

A number of studies have been submitted to regulators considering whether particular network elements are essential for competition (see Tardiff, 1998b; and Huber and Leo, 1999). The data presented in these studies demonstrate competitive alternatives are present for the following elements; consequently, there is no economic basis for mandatory unbundling of them in markets where these competitive alternatives exist or are economically feasible.[29]

(a) *Local loops provided to medium and large businesses, particularly in metropolitan areas.* Facilities-based CLECs are serving these customers and can expand in a timely fashion to serve nearby customers. Huber and Leo report that as of early 1999 CLEC facilities already served 15 per cent of all commercial buildings and between 8 per cent and 18 per cent of the business customers they have targeted.

(b) *Switching.* The growth in CLEC switches, the fact that CLEC switches tend to serve larger areas than ILEC switches, and the availability of technological alternatives to the typical ILEC switch demonstrate that CLECs do not require switching inputs from ILECs.

(c) *Transport.* CLECs generally co-locate their networks in larger ILEC central offices and in these areas CLECs generally tend to have their own fiber facilities. Therefore, transport is clearly not essential is such areas.

(d) *Operator services and directory assistance.* Numerous providers are offering operator services and directory assistance in direct competition with ILEC offerings. In some cases ILEC subsidiaries acquire such services from competitive sources rather than the affiliated ILEC.

(ii) Pricing considerations

The proponents of the existing forward-looking cost models argue that an ILEC offering only unbundled elements on a wholesale basis would be fully compensated, because an efficient firm could sell elements to retail providers at these prices while at the same time recover its costs and earn an adequate return on its investment. Accordingly, when ordering such prices, the regulator behaves like a central planner in the sense that the prices predicted from the model are believed to be a close approximation of the outcome of the competitive process. That is, in the process of attempting to mimic a competitive outcome, use of model-generated prices substitutes the long-run forecast implicit in the model for competition itself.

Another issue is whether the efficient firm models should be used to set the prices of all network elements required to be unbundled or whether such regulatorily-set prices should be limited only to those elements that are necessary for competition and are supplied on a monopoly basis by the ILEC – essential facilities. For example, the FCC's rules provide an attractive range of inputs. Should all such inputs be priced on an efficient firm basis? The answer to this question follows from the definition of essential facilities. When an element is not an essential facility then economic alternatives exist and competition for market elements is feasible.[30] In this case economic efficiency is best provided by market outcomes rather than regulatory price setting. In particular, allowing competition to replace regulation in the case of network elements that are not essential facilities would be best facilitated by the use of negotiated rates (with non-discrimination provisions to make available contract terms to similarly situated CLECs) in place of regulatorily imposed prices.

Implications for the Pricing of ILEC Retail Services

The case of network elements that remain essential facilities imposes some restrictions on the incumbents' competitive retail prices. In particular, those retail prices must satisfy an imputation requirement that they should include at least as much contribution (margin over cost) as that contained in the essential facilities used by competitors (Hausman and Tardiff, 1995b). This requirement is equivalent to the test for predatory pricing when no essential facilities are present – that retail prices exceed incremental cost. Competition for retail services will potentially involve the ILECs competing against providers using some of the ILEC's network elements as inputs as well as other providers using their own facilities almost exclusively. These prospects raise the question of whether the ILEC's retail prices should be restricted by regulation. In particular, should the prices that ILECs charge for unbundled network elements (UNE) affect the prices for their competitive retail services?

Competing approaches to establishing a retail price floor have emerged. The first is essentially a 'sum of the UNE price' rule for retail price floors. That is, in order for CLECs that provide retail service by purchasing UNEs from the ILEC to have sufficient margin to compete, the ILEC's retail prices are required to impute the UNE prices. While this rule perhaps balances competition between the ILEC and CLECs pursuing a UNE strategy, it overlooks the possibility that the ILEC could be artificially disadvantaged in competing against facilities-based rivals. The reason is that to the extent that the UNE prices include mark-ups above incremental cost, imputation of the UNE prices produces the result that the ILEC retail prices cannot be lowered all the way to incremental cost. In contrast, facilities-based CLECs face no such constraints. Therefore, the 'sum of the UNEs rule' protects a certain type of competitor – one that adds its own retail functions on to inputs provided by the ILEC – but ignores the effects on competition for those wholesale functions themselves. Consequently, the alternative rule would require imputation of only the mark-ups over cost contained in network elements deemed to be essential facilities. For network elements made available either on a voluntary basis or because regulators have mandated certain non-essential elements to be available, perhaps for a limited period, the ILEC's retail price floor would not contain mark-ups for such elements.[31] Two recent regulatory decisions (CRTC, 1997; California Public Utilities Commission, 1999) have selected this latter price floor rule.

The difference between the rules is illustrated by an example. Assume residential LEC service is provided with loop, switching and retail function inputs. Table 7.1 lists the prices and costs of the elements. Here the loop and switching are unbundled, but the retail functions are not, because any CLEC is able to provide them. Under the 'sum of the UNE rule', the minimum price the ILEC is allowed to charge for its retail service is USD26 (USD18 + USD5 + USD3). The rationale for this rule is to allow CLECs that purchase loops and switching from the ILEC, but provide their own retail functions, to be able to charge a lower price when they provide retail functions more efficiently. However, this rationale breaks down in the case of facilities-based competition. For example, suppose the cable television company can provide telephone service at a cost of USD23. This cost is higher than the ILEC's (USD15 + USD4 + USD3 = USD22), but lower than the price floor. Such a CLEC could undercut the ILEC (and also the CLEC using UNEs), even though its costs are higher.

Table 7.1　Residential local service prices and costs (USD)

Element	Price	Cost
Loop	18	15
Switching	5	4
Retail	n.a.	3

NOTES

1. While the volume of data traffic has surpassed voice volumes, the latter still generate 80 per cent to 90 per cent of revenues, due to lower per-unit prices for data.
2. For example, the Federal Communications Commission (FCC, 1996) developed its version of total-service long-run incremental cost for the purpose of pricing unbundled network elements, such as subscriber lines. The FCC named its approach 'total element long-run incremental cost' or TELRIC.
3. These mark-ups tend to be uniform not only over the entire volume of production, but also over time. For example, if the equipment used to provide subscriber lines had an economic life of ten years, total service cost calculations typically produce uniform prices that would need to be charged for the full ten years in order to allow the firm to recover its costs.
4. For example, television viewers are bombarded with a plethora of long distance options. These include AT&T's one rate plans, where for a fixed monthly fee a subscriber receives a single price for calls made anywhere covered by the plan at any time, plans that have lower rates for off-peak periods, plans that offer the same price for calls up to 20 minutes, and a new plan by Sprint that offers up to 1000 minutes of calling for a flat monthly price. On top of all these, are Internet offerings that have even lower prices.
5. In telecommunications, carriers clearly target different types of customers. While the incumbent long-distance carriers have provided their best offers to higher volume users (see Kahn et al., 1999), incumbent local exchange carriers have served low-volume users to a much larger extent when they have been permitted to provide long-haul service. For example, SNET serves 50 per cent of residential customers in Connecticut but captures only 15 per cent of revenues (Murray, 2000). Similarly, Bell Atlantic's (now Verizon) new long-distance services are particularly attractive to low-volume users in New York (*Telecommunications Reports*, 2000).
6. Since 1936, the inflation in telecommunications prices has been about 2 per cent per annum lower than the inflation in the consumer price index in the US (FCC, 2000b).
7. The implications of technological and competitive developments for telecommunications costs and prices are dealt with extensively in Alleman and Noam (1999).
8. A possible explanation for the large aversion to up-front charges in the case of ordinary telephone service is the fact that some customers tend to move frequently, thus shortening the period over which the initial charge is effectively amortized.
9. I have examined two proprietary wireless studies conducted in the early and mid-1990s, both of which indicated that a high up-front charge was more onerous than ongoing monthly charges. The behavior of suppliers is completely consistent with this consumer preference – the price of the wireless telephone is typically discounted substantially. In the case of wireless services, the desire to avoid large up-front charges is consistent with the real options literature. Because of the typically large price reductions in electronic services, such as wireless telephones and/or the desire to change providers, the discount on the telephone price might be compensation to the consumer equivalent to the value of keeping one's options open.
10. Call waiting is the most successful of telephone companies' traditional custom calling services, with subscribership of the 30 per cent to 40 per cent in the US.
11. This section and the next are based on Kahn and Tardiff (2000).
12. In a book devoted to the proposition that vigorous enforcement of the antitrust laws is necessary for the preservation of fair competition, Dirlam and Kahn (1954) began the chapter 'Business Integration and Monopoly' with the proposition: 'competition requires . . . that business units be free, ordinarily, to take on new products, new functions, or enter new markets – in short, to integrate'.
   And, in summarizing the problem of applying the antitrust laws to the operations of integrated firms, they observed:

The perplexing problem is that the competitive advantages stemming from gains in efficiency attributable to integration are in practice inseparable from the merely strategic advantages that pose the dangers to society just described. Efficiency gains arise from the fuller utilization of a firm's capacity, whether measured by its physical plant, managerial talents, by-products, technological skills, or the ideas issuing from its research laboratories.

13. Similar observations are made by de Chazeau and Kahn (1959, p. 261) in Kahn (1988, Vol. II, pp. 261–2).
14. While the October 2000 announcement by AT&T that it intends to spin off its wireless and cable operations from its core long-distance and network services is clearly a change in the strong trend towards increasing vertical integration discussed in detail below, it does not invalidate the general proposition that telecommunications firms face the need to integrate across formerly separate markets. Indeed, even AT&T's separated companies will be quite integrated by historical standards. For example, its cable operations now offer packages of local telephone, long-distance, cable television, and high-speed Internet access and would probably continue these offerings on becoming a separate company. The fundamental point is that companies appear to be exploring ways of combining operations and offering attractive product offerings. That some of these explorations do not come to fruition is not surprising in the light of the very dynamic and fluid nature of the industry, the regulatory constraints that still persist, and the apparently short-term perspective of investors that seems to demand financial results rather quickly. Indeed, some have observed that AT&T's investments to vertically integrate still make sense, but that investors do not have the patience to see them to fruition, especially in light of the recent downturn in telecommunications stocks in general, and AT&T's in particular.
15. In 1997, AT&T announced a new wireless system that would link customers directly to its network (Keller, 1997) along with its expectation of introducing that technology within three to four years (*Telecommunications Reports*, 1998). Keller described the system as 'a small transceiver mounted on the side of a house, could give AT&T lightning-fast entry into the local phone business'. Gregory Rosston (RCR, 1997) then the FCC Deputy Chief Economist commented that, 'AT&T recently announced that it is developing PCS technology to be what Chairman Reed Hundt has termed the "Raiders of the Local Loop".' The fact that AT&T seems recently to have shifted the focus of its competitive strategies to the use of cable provides a further, even more dramatic illustration of the kaleidoscopic variety of new ventures and blurring of boundaries – and the consequent absence of necessity, indeed counterproductiveness of continued regulatory restrictions on entry.
16. At the same time, AT&T was exploring alliances with other cable television providers, such as Time-Warner, with the intention of extending its ability to serve residential customers with its own local facilities. AT&T Chairman Michael Armstrong, striking a defiant stance, said cable television companies that do not partner with the long-distance company to offer local telephone service still deal with AT&T but as a rival (Cauley, 1998). In further confirmation of the seriousness of its intentions, Blumenstein and Cauley (1998) report that,

> in an unusual move, AT&T Corp. has committed to buy as much as USD900 million of equipment to deliver telephone service over Tele-Communications Inc.'s cable-TV lines, even though the phone company's landmark purchase of the cable giant hasn't been completed. . . . The contract is part of what is expected to be a multibillion-dollar investment by AT&T to ready TCI's cable lines to deliver voice, Internet and interactive video.

17. Robert Allen, AT&T's then chairman, stated on 8 February, the day the 1996 Telecommunications Act became law, said that it had the ability to connect directly with its large business customers to offer local exchange service. It helps to put the 100

switches into perspective to point out that the RBOCs currently have about 4600 switches (not including remotes). On the other hand, because the switches of new local exchange entrants are likely to be placed in areas with higher volumes and such entrants will be able to obtain unbundled switching from the ILECs, this simple comparison of the respective numbers understates their importance. The 100 – or, adding Teleport's, 150 – are still, of course, only a small percentage of the approximately 4600 owned by the RBOCs collectively. On the other hand, that account understates their relative importance, because the switches of the new local exchange entrants, including AT&T, tend to be placed in areas with higher volumes. Moreover, they of course understate the ability of an AT&T to compete for local service customers, because that company and the other entrants are able to obtain unbundled switching from the ILECs as well.

18. MCI WorldCom's 1998 Annual Report describes annual local revenues of about USD400 million and an annual growth rate of 80 per cent.

19. March 1999 figures are based on the Bellcore *Local Exchange Routing Guide*.

20. For example, SCANA Corp., the parent company of South Carolina Gas and Electric, controlled 2500 route miles of cable fiber back in 1995 through its subsidiary MPX Systems, Inc., and was planning to double that (*Fiber Optics News*, 1995).

21. Calculated from data reported in the FCC's *Statistics of Communications Common Carriers*.

22. ALTS (2000) report that CLEC capital expenditures had grown to USD15 billion in 1999.

23. At the end of 1999, cable modems served 71 per cent of the 1.7 million customers using high-speed access and DSL served 29 per cent.

24. Even though voice CLECs have made a smaller dent into the incumbents' volumes, the FCC (2000a) observed that local competition is farther along than was long-distance competition at a comparable stage.

25. Weisman (2000) noted that the forward-looking cost of the firm in question has always been the proper standard for a predatory pricing test. He further noted the contradiction between forcing the incumbent's retail prices to be at or above this traditional standard while at the same time requiring that it sell certain essential inputs to competitors based on a lower cost efficient firm standard. In effect, the incumbent would be foreclosed from the retail market by such contradictory cost standards.

26. US WEST's proposal for funding investments, rather than a putatively equivalent monthly subsidy for universal service is equivalent to the use of long-term contracts for unbundled network elements (Kahn and Tardiff, 1997).

27. The magnitude, but not the existence, of such economies has become an issue of some controversy, especially in the context of the FCC's TELRIC standard. Proponents of this standard argue that costs that might be shared among services become attributable to particular elements. For example, certain fixed costs associated with local switching are shared among services using the switch, for example local and toll calling, but are direct costs when switching is sold as an unbundled element. While this viewpoint may have some merit when the ILEC is only providing unbundled elements on a wholesale basis and there are no alternative sources for these elements, it becomes problematic when the incumbent is a vertically integrated provider of wholesale and retail services and/or there are alternative providers of network elements.

28. Economies of scope exist in long-distance markets with prices set much closer to cost for large users reflecting the more intensive competition for these customers.

29. The FCC (1999) did, in fact, remove from the list of mandatory unbundled elements directory and operator services and switching used to serve business customers with four or more lines in certain metropolitan areas.

30. That some network elements are not essential facilities is of more than academic interest. For the major components of the network – subscriber loops, switching, and interoffice transport – the CRTC (1997) determined that only subscriber loops in lower density areas are essential facilities, because competitive switching and transport alternatives are available.

31. For example, the CRTC (1997) generally limited mandatory unbundling to essential

facilities. However, certain non-essential elements, such as loops in high-density areas and transport, are required to be unbundled for a limited duration in order to speed up the development of competition.

REFERENCES

Alleman, J. and Noam, E. (eds) (1999), *The New Investment Theory of Real Options and its Implications for Telecommunications Economics*, Boston: Kluwer.

Association for Telecommunications Services (ALTS) (2000), 'The state of competition in the US local telecommunications marketplace', February.

AT&T (1998), 'AT&T completes TCG merger; TCG now core of AT&T local services network unit', News Release, 23 July.

AT&T and TCI (1998), 'AT&T, TCI to merge, create new consumer services unit', News Release, 24 June, http://www.att.com/press/0698/980624.cha.html.

Blumenstein, R. and Cauley, L. (1998), 'AT&T agrees to purchase equipment to deliver phone service on TCI's lines', *Wall Street Journal*, 30 October, p. B2.

California Public Utilities Commission (1999), 'Interim decision setting final prices for network elements offered by Pacific Bell', Decision 99-11-050, 18 November.

Canadian Radio-television and Telecommunications Commission (CRTC) (1997), 'Local competition decision', Telecom Decision CRTC 97–8, Ottawa, 1 May.

Cauley, L. (1998), 'AT&T chairman presses cable firms on cable ventures', *Wall Street Journal*, 3 November, p. B19.

Council of Economic Advisors (1999), 'Progress report: Growth and competition in US telecommunications 1993–1998', 8 February.

de Chazeau, M.G. and Kahn, A.E. (1959), *Integration and Competition in the Petroleum Industry*, New Haven: Yale University Press, p. 261.

Dirlam, J.B. and Kahn, A.E. (1954), *Fair Competition: The Law and Economics of Antitrust Policy*, Ithaca, NY: Cornell University Press (reprinted by Greenwood Press, 1970), pp. 150–51.

Ebbers, B.J. (1999), 'MCI–Sprint is no big deal', *Wall Street Journal*, 8 October, p. A18.

Engebreston, J. (1997), 'The new guys in town', *Telephony*, 2 June, pp. 98–110.

FCC (1996), *Implementation of the Local Competition Provisions in the Telecommunications Act of 1996*, CC Docket No. 96-98, First Report and Order (adopted 1 August, released 8 August).

FCC (1999), *Implementation of the Local Competition Provisions in the Telecommunications Act of 1996*, CC Docket No. 96-98, Third Report and Order and Fourth Further Notice of Proposed Rulemaking (adopted 15 September, released 5 November).

FCC (2000a), *Telecommunications @ the Millennium: The Telecom Act Turns Four*, Office of Plans and Policy, 8 February.

FCC (2000b), *Trends in Telephone Service* (March), Table 13.1.

Fiber Optics News (1995), 'Growing utility fiber market tempered by considerable hesitancy', 15(19), 15 May.

Gold, H. (1998), Statement to the Federal Communication Commission's *En Banc* on the State of Local Competition, 29 January.

Hausman, J.A. (1997). 'Valuation and the effect of regulation on new services in telecommunications', *Brookings Papers on Economic Activity, Microeconomics*: 1–38.

Hausman, J.A. (1999), 'The effect of sunk costs in telecommunications regulation', in Alleman, J. and Noam, E. (eds), *The New Investment Theory of Real Options and its Implications for Telecommunications Economics*, Boston: Kluwer.

Hausman, J.A. and Tardiff, T.J. (1995a), 'Benefits and costs of vertical integration of basic and enhanced telecommunications services', prepared for filing with the Federal Communications Commission, Computer III Further Remand Proceedings, CC Docket No. 95-20, on behalf of Bell Atlantic, Bell South, NYNEX, Pacific Bell, Southwestern Bell, and US West, 6 April.

Hausman, J.A. and Tardiff, T.J. (1995b), 'Efficient local exchange competition', *Antitrust Bulletin*, 40, 529–56.

Hausman, J., Tardiff, T. and Belinfante, A. (1993), 'The effects of the breakup of AT&T on telephone penetration in the United States', *American Economic Review*, 83, 178–84.

Huber, P.E. and Leo, E.T. (1999), 'UNE fact report', submitted to the Federal Communications Commission by USTA on behalf of Ameritech, Bell Atlantic, BellSouth, GTE, SBC, US WEST, CC Docket No. 96-98, 26 May.

Kahn, A.E. (1988), *The Economics of Regulation*, Cambridge, MA: MIT Press.

Kahn, A.E. and Tardiff, T.J. (1997), 'Funding and distributing the universal service subsidy', prepared for US West for presentation to the Federal Communications Commission, 13 March.

Kahn, A.E. and Tardiff, T.J. (2000), Public interest affidavit before the Federal Communications Commission in the matter of application of SBC Communications, Inc., Southwestern Bell Telephone Company, and South-western Bell Communications Services, Inc., d/b/a Southwestern Bell Long Distance for Provision of In-Region, InterLATA Services in Texas, on behalf of Southwestern Bell Telephone Company, filed 10 January.

Kahn, A.E., Tardiff, T.J. and Weisman, D.L. (1999), 'The Telecommunications Act at three years: An economic evaluation of its implementation by the Federal Communications Commission', *Information Economics and Policy*, 11, 319–65.

Keller, J.J. (1997), 'AT&T unveils new wireless system linking home phones to its network', *Wall Street Journal*, 26 February, p. B4.

Lenahan, G. (1999), 'Growth of voice and data traffic in US public networks', Telcordia Technologies.

Mandy, D. (2000), 'TELRIC pricing with vintage capital', presented at the Advanced Workshop in Regulation and Competition, 13th Annual Western Conference, Center for Research in Regulated Industries, Rutgers University, Monterey, California, 6 July.

Murray, S. (2000), Public interest affidavit before the Federal Communications Commission in the matter of application of SBC Communications, Inc., Southwestern Bell Telephone Company, and Southwestern Bell Com-munications Services, Inc., d/b/a Southwestern Bell Long Distance for Provision of In-Region, InterLATA Services in Texas, on behalf of Southwestern Bell Telephone Company, filed 10 January.

New Paradigm Resources Group (1998), *Review of the Annual Report on Local Telecommunications Competition*, March, p. 2.

Perl, L.J. (1984), 'Residential demand for telephone service: Preliminary results of a new model', in Mann, P.C. and Trebing, H.M. (eds) *Changing Patterns in Regulation, Markets, and Technology: The Effect of Public Utility Pricing*, Proceedings of the Institute of Public Utilities Fifteenth Annual Conference, Michigan State University.

RCR, (1997), 3 March, p. 59.

Schiesel, S. (1998), 'FCC blocks two Bells on long-distance entry', *New York Times*, 29 September.

Shapiro, C. and Varian, H.R. (1999), *Information Rules*, Boston: Harvard Business School Press.

Smith, A.T. (1994), Southern Bell testimony before the Florida Public Service Commission, Docket No. 930330-TP, 1 July.

Tardiff, T.J. (1995), 'Effects of presubscription and other attributes on long-distance carrier choice', *Information Economics and Policy*, 7, 353–66.

Tardiff, T.J. (1998a), 'Pricing and new product options with telecommunications competition', in Dolk, D.R. (ed.) *Proceedings of the Thirty-first Annual Hawaii International Conference on Systems Sciences, Vol. V, Modeling Technologies and Intelligent Systems Track*, Los Alamitos: IEEE Computer Society, 6–9 January, pp. 416–25.

Tardiff, T.J. (1998b), Testimony before the California Public Utilities Commission, on behalf of Pacific Bell, 8 April.

Tardiff, T.J. (1999), 'The forecasting implications of telecommunications cost models', in Alleman, J. and Noam, E. (eds) *The New Investment Theory of Real Options and its Implications for Telecommunications Economics*, Boston: Kluwer.

Telecommunications Reports (1996), 'Study says consumers would buy bundled services', 12 August.

Telecommunications Reports (1998), 'Zeglis says lower access charges spurred TCG merger', 4 May, p. 33.

Telecommunications Reports (2000), 'Low-volume users may be biggest winners from Bell Atlantic's long distance market entry', 10 January, p. 3.

Train, K. (1994), 'Self-selecting tariffs under pure preferences among tariffs', *Journal of Regulatory Economics*, 6, 247–64.

Wall Street Journal (1995), 'AT&T vows battle to offer local service', 27 October, p. A4.

Weisman, D.L. (2000), 'The (in)efficiency of the "efficient firm" cost standard', *Antitrust Bulletin*, 45, 195–211.

8. Mobile telecommunications and regulatory frameworks

Harald Gruber and Tommaso M. Valletti

INTRODUCTION

For the last 20 years mobile telephony and Internet technology have dramatically changed the telecommunications sector. Mobile communications started as a premium service offering voice transmission with mobility. As the service became more common, mobile telephony challenged the notion of natural monopoly within the sector and unravelled a wave of regulatory change that has deeply changed the market structure of the telecommunications industry (Regli, 1997; Laffont and Tirole, 2000). In many countries there are now more mobile telephones than fixed lines. These changes were preceded by profound technological advances, particularly in microelectronics, which provided a platform for the development of cellular mobile services. However, the key to industry development was the better use of radio spectrum, a scarce resource with alternative uses. Industry evolution can be seen as a race to relax this constraint through increased spectrum efficiency – that is, by employing smaller amounts of spectrum per unit information transmitted – and by political lobbying for more frequency. Relaxation of the spectrum constraint has allowed entry, however, entry remains restricted by regulation. The resultant market structure is oligopolistic. On the basis of cross-country evidence it is possible to assess the empirical validity of a range of oligopoly models. This research is relevant for industrial policy since it provides useful information for the future design of market structure and sector regulation. Oligopoly rents are pervasive, as the high profitability of the established firms in the sector show, and even expected profitability is substantial as illustrated by the proceeds from the auctions for third generation (3G) services in Europe. Auctioning spectrum for mobile services has attracted much public interest, and has been a rewarding application of game theory in selecting the most efficient operators and in extracting oligopoly rents.

Mobile telecommunications is a network industry and network effects manifest themselves in very particular ways. Several reciprocally incompatible mobile standards have been adopted globally. Network effects are due to economies of scale in the production of equipment, in the operation of the networks and, for users, in the ability to roam; that is, the use of handsets in areas other than those covered by the firm the user subscribes to. Here, also, the industry offers an interesting test field for theoretical propositions taken from the literature on standards. An aim of this chapter is to illustrate how the interplay of technological development and regulatory change has shaped the evolution of the mobile telecommunications industry. The role of government is fundamental in setting the framework for the sector, in particular by establishing the technological base (voluntary versus compulsory standards), the number of firms and their mode of entry. Investigating the determinants of market structure in industries with large fixed (sunk) costs is a prominent subject of the IO research agenda, and this industry represents an attractive laboratory (Sutton, 1998). There is also a wide range of other interesting research questions which have a broader application beyond mobile telephony, but that can be tackled with the data at hand for this industry. For instance, the sector has developed business strategies on pricing and product differentiation. Market growth is spectacular and panel data are available for looking at the diffusion of technology. This chapter makes a survey of the most prominent empirical research, and makes suggestions on fields that deserve further work. The following section briefly presents the technological evolution of the industry in the race to relax its main limiting resource: radio spectrum. Demand issues and supply aspects (such as firm behaviour and business strategy) are discussed, respectively. The penultimate section describes in detail the important role of regulation for this industry. A final section concludes the argument.

TECHNOLOGY

Mobile telecommunications use radio waves instead of wires to connect users. The available portion of the radio frequency suitable for mobile use in the overall spectrum is limited both by technology and regulation. The earliest applications for mobile communications date back to the 1920s.[1] Since there are alternative uses for the spectrum, such as broadcasting and military applications, firms struggled to convince governments to allocate a substantial portion of it to mobile services (Kargman, 1978; Levin, 1971).[2] An important technical improvement occurred during the 1960s with the development of the cellular concept of mobile telecommunications. The system is called cellular because the area to be served is divided into cells with an antenna in the middle.[3] This design, coupled with sophisticated electronics, allowed for the more efficient use of spectrum and the accommodation of a much larger number of users for a given frequency range. But mainly because of regulatory delay, cellular systems were only deployed at the beginning of the 1980s (Calhoun, 1988). Cellular technology can be distinguished according to the way in which the signals are transmitted: analogue (first generation) and digital (second generation) technology starting at the beginning of the 1990s. Because of network effects, only a limited number of competing systems were developed: seven analogue and four digital systems found application. These systems differ by a number of technological parameters, and in their ability to use the spectrum efficiently. Table 8.1 lists the systems indicating the country and year of first adoption, and the efficiency parameter.

Analogue Systems (First Generation Mobile)

The large number of analogue systems in the early cellular industry occurred because cellular communications were viewed as just a new business of

Table 8.1   Characteristics of cellular systems

System	Country	First adoption	Spectral efficiency (bits/second/Hertz)
Analogue			
NTT	Japan	1979	0.012
NMT-450	Scandinavia	1981	0.048
NMT-900	Scandinavia	1986	0.096
AMPS	US	1983	0.333
TACS	UK	1985	0.320
C-450	Germany	1985	0.264
RMTS*	Italy	1985	n.a.
RC 2000*	France	1985	n.a.
Digital			
GSM 900	EU	1990	1.35
GSM 1800	EU	1993	1.35
DAMPS	US	1991	1.62
IS-95 (CDMA)	US	1993	1.75**
PDC	Japan	1993	1.68

Notes:
* Also called quasi-cellular because of restrictions on handover between cells.
** Not strictly comparable.

Source: Authors, based on Garg and Wilkes (1996) and Rappaport (1996).

state-owned monopoly operators. Thus the development of the cellular network also provided a means of honing the innovative capability of national equipment suppliers (ITC, 1993; Funk, 1998). Figure 8.1 illustrates the difference in popularity between analogue systems by indicating the number of national cellular networks that have been adopted for the systems. The AMPS system, which was developed in the US and adopted as the national analogue standard, had the largest number of national networks. On the basis of spectral efficiency it also ranks first for analogue systems. The NMT system, developed in Scandinavia, is one of the least efficient systems but attained the largest number of networks in Europe and ranks second worldwide because of its early entry.[4] The TACS system, a modified version of the AMPS system to cater for European spectrum allocations, ranks third

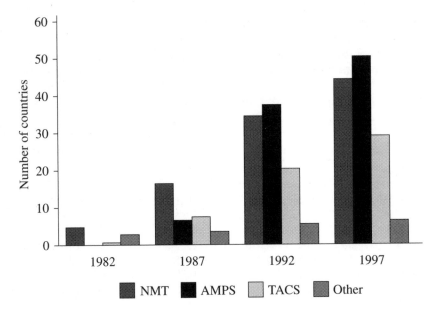

Figure 8.1   Adoption of analogue cellular technology

in terms of number of networks. The C-450 system, which is technically quite sophisticated and efficient, was adopted in only two countries besides Germany. The remaining analogue cellular systems (NTT, RC2000 and RTMS) were not adopted outside their country of development – that is, Japan, France and Italy respectively.[5] This comparison shows that the most successful systems (in terms of the number of adopting countries) emerge when the domestic market is sufficiently large (AMPS for the US) or governments coordinate with other countries (NMT for Scandinavia). The large diffusion of TACS follows from it being an adaptation of AMPS to another frequency band. NTT, C450, RTMS and RC2000 show a national market alone was usually too small to economically support the development of additional incompatible systems. Thus network effects and economies of scale have led to the emergence of multiple designs for analogue cellular: AMPS (and its derivative TACS) and NMT.

Digital Systems (Second Generation Mobile)

With four systems worldwide there are fewer digital cellular systems than analogue systems. This is because European countries adopted GSM as a common standard,[6] whereas non-European countries such as the US permitted competing digital systems.[7] DAMPS was introduced in the US to ensure smooth transition of the prevailing analogue standard to a digital system. The systems were made compatible by dividing an analogue channel and thereby tripling capacity. DAMPS has greater spectral efficiency than GSM. IS-95 CDMA is an alternative digital system not compatible with DAMPS. Performance comparisons between systems are difficult. However, a widely shared guess of the capacity advantage of CDMA over GSM is about 30 per cent (Garg and Wilkes, 1996; Webb, 1998). JDC was introduced as a national standard in Japan to establish this system as a regional standard under the name Pacific Digital Cellular, but this attempt failed as no country outside Japan adopted the system.

The digital technology spectrum efficiency is about three to six times that of analogue technology. Digital technology has advantages including less noise in operation and enhanced privacy. Figure 8.2 shows the differences in popularity between digital systems. GSM was the first to be introduced on a large scale and has since dominated the market. In 1998, of the 193 million digital subscribers worldwide, more than 72 per cent were GSM subscribers (ITU, 1999). A reason for GSM's success is that network effects are pervasive and the first-mover advantage outweighs any disadvantage in technological efficiency.

Digital Systems (Third Generation Mobile)

Whereas the first and second generations (2G) of mobile telecommunications systems were mainly designed for voice transmission, the next technological step was the development of systems for data transmission. 3G systems are being developed that will dramatically increase data rates by a factor of at least 40. The International Telecommunication Union (ITU) is trying to achieve a global standard for 3G mobile telecommunications through its initiative IMT-2000. This should be characterized by seamless

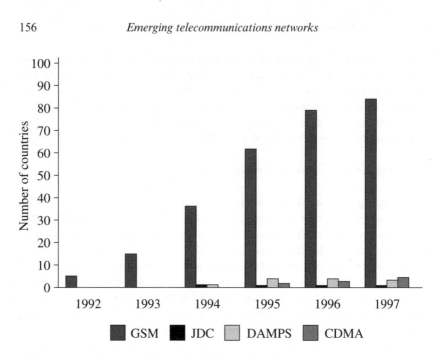

Figure 8.2   Adoption of digital cellular technology

global roaming to enable users to move across borders using the same number and handset, and at least a 40 times higher signal transmission rate than 2G systems, allowing for fast Internet access. ITU has accepted five systems for the family of IMT-2000 standards that satisfy technical requirements to provide 3G services – three based on CDMA and two on TDMA. Only W-CDMA (also known as UMTS and is promoted by ETSI, the European standard body), CDMA2000 (the evolution of CDMAOne technology promoted by the US CDMA patent holder Qualcomm) and IWC (based on TDMA and compatible with DAMPS) are expected to find widespread adoption. Failure to reach an agreement on a 3G standard gives rise to the need for multiple mode and band handsets capable of handling alternative modes and frequency bands. This would enable worldwide roaming but at a higher cost than with a single standard, due to the increased complexity of handsets and networks. The failure to agree on a world standard is also due to the fact that mobile firms seek backward compatibility for their installed mobile systems.

In Europe 3G systems are referred to as Universal Mobile Telecommunications Systems (UMTS), a concept developed by the ETSI and based on CDMA technology. European interest was in making UMTS as much as possible backward compatible with the existing GSM which is based on TDMA.[8] This objective is in conflict with making it compatible with CDMA2000. The first adoption of 3G systems should occur from 2002 in Europe and Japan. The US are delaying the introduction of 3G systems, mainly because of the slow development of the 2G systems that have been launched late and using a range of different, non-compatible technologies (ITU, 1999). European policy makers are keen to introduce 3G early. Early adoption of UMTS is seen as key to preserving the world-wide lead in mobile telecommunications technology established with GSM (see European Commission, 1997). Gruber and Hoenicke (2000) elaborate on the question of whether the speed of adoption proposed and the size of required investment are warranted by a sufficiently high demand. A simulation exercise shows that revenues from data services have to increase substantially to make UMTS a profitable undertaking.

Data transmission rates of 2G systems, which were limited due to low speeds (9.6 kbit/s in GSM), are being enhanced by features such as HSCSD (High Speed Circuit Switched Data), GPRS (General Packet Radio Service) and EDGE (Enhanced Data Rates for GSM Evolution) that allow several times higher data rates (about 100 kbit/s). While these steps towards higher data capacity (referred to as 2.5 generation) constitute forerunners to 3G-telephony and may pre-empt or delay its introduction, 2G systems can now provide data services. For instance, WAP (Wireless Application Protocol) is a specification for a set of communication protocols to standardize the way that mobile telephones can be used for Internet access. Although Internet access has been possible in the past, different manufacturers have used different technology. In the future devices and service systems that use WAP will be able to interoperate.[9]

DEMAND SIDE

Since its introduction in the mid-1980s mobile communications adoption has grown rapidly. In the OECD area at June 1999 the number of cellular mobile subscribers reached 300 million. Between 1992 and 1997 the number of mobile subscribers increased at a compound annual growth rate of 52 per cent, implying the customer base doubled every 20 months. Penetration rate was 1 per cent on average in 1990 and increased to 27 per cent by June 1999. Nevertheless, there are still enormous differences in national penetration rates. For instance, in Finland there were 61 mobile subscribers per 100 inhabitants in 1999, while Mexico had just five per 100 inhabitants (OECD, 2000). Despite its tremendous success relatively little is known about the determinants of demand. Hausman (1997) estimates a price elasticity of −0.51 for the US, which is more inelastic than long-distance services but more elastic than other telecommunications services. While it may have been reasonable to consider average profiles in early network development, when most subscribers were heavy business users, now such averaging is not satisfactory because it neglects the growing heterogeneity of consumer preferences.

In addition to the usual market segments – business and residential – mobile telephony involves aspects related to the underlying technology. In particular, there is a clear relationship between network coverage (including roaming capability) and user mobility. Cell capacity is related to usage and reliability.[10] Consumer tastes differ in terms of mobility, usage pattern and congestion. Users range from business executives travelling internationally making many calls to self-employed persons confined to a geographical area and making few telephone calls. Other subscribers may need telephony for emergency use only (option value). Customers who do not want their calls dropped must be prepared to pay a premium while others will not. These examples illustrate that international roaming, network coverage and network reliability are variables that play an important role in the strategic interaction among operators.

A recent service that has proven highly popular among mobile users is the short message service (SMS); that is, the ability to send and receive text messages on mobile handsets. While demand for 3G services is unproven, the success of SMS suggests that this may become a fast growing area for mobile data. Other key areas promise to be financial (such as banking, credit card, financial news and trading) and mobility (such as hotel booking, weather and traffic information) services. As mobile telecommunications services are still relatively new, first-time users are not perfectly informed about their use. Entry costs, like handset price, play a major role in a customers decision to subscribe to the network, as does ignorance of likely usage. A seller might deliberately use this knowledge and introduce price discrimination.[11] While over time misinformation among mobile users has decreased the phenomenon is still important among fixed telephony customers calling mobile users.

While new empirical work is needed to understand the adoption and usage choices, some results can be obtained through another avenue of research. If market adoption follows an epidemic model whereby the flow of adopters is exogenously related to the stock of existing adopters, then diffusion typically follows an S-shaped function.[12] Gruber and Verboven (2000a) follow this approach by exploiting the cross-sectional variation of diffusion rates in EU countries for the period 1984 to 1997. They show the growth of mobile telephony was mainly driven by the introduction of digital technology that brought an expansion of capacity. An increase in competition also enhances growth, however, digitalization had a much stronger impact. The study also finds weak evidence of substitutability between fixed and mobile telephony. Interestingly, wealth of the population (as measured by GDP) is not crucial in explaining the diffusion of mobile services – contrary to many claims – and that diffusion levels should converge over time. This phenomenon, whereby lagging countries catch up, is also found by Gruber (2001) for Central and Eastern Europe (CEE). While technology has smaller explanatory power than in EU countries, diffusion in CEE is enhanced by long waiting lists for fixed telephones, higher levels of urbanization and the size of the fixed network.

SUPPLY SIDE

The mobile communications industry is a rich laboratory for the study of oligopoly theory. Gradual relaxation of the spectrum constraint has allowed the release of additional licences, which has resulted in changing market structures. At the OECD level three-quarters of countries with mobile services were monopolies at 1990. By 1995, one-third of the adopting countries had monopoly providers, half had duopoly providers and the remainder had more than two operators. At 2000, of the 29 OECD countries with mobile communication networks none had a monopoly provider, four were serviced by duopoly providers, 11 had three operators and 14 had more than three operators offering services (OECD, 2000).

Pricing

Typical patterns in the evolution of pricing strategy are readily identified (ITU, 1999).[13] The early stage of mobile market development, roughly the 1980s, was led by business demand. Users were prepared to pay high prices for service and this resulted in spectacular profit margins. Leading monopoly and duopoly firms did not adopt new strategies for the expansion of personal communications until they faced increasingly competitive markets. Pricing was uniform with no flexibility for users with different usage patterns. Pricing strategies mostly were designed to ration the available spectrum capacity, for example charging higher prices in densely populated areas.

In the early 1990s operators started to address the needs of new types of users, such as mobile professionals, the self-employed and salespeople, with a wider range of tariff packages. As most market structures were duopolistic there was little incentive for mobile operators to cut prices in the face of demand growth. Rather they tended to deviate from fixed-line pricing policy with mobile telephony prices not being distance sensitive. Entrants forced

incumbents to be more responsive and growth was driven by product differentiation rather than price reductions.[14] This process of price differentiation continued until the mid-1990s when the residential market was
targeted. There has since been a proliferation of tariff packages due to entry
as new capacity became available. Since operators cannot perfectly discriminate among consumers they have been careful when designing tariffs to take
into account the effects on existing subscribers.[15] From the late 1990s mobile
telecommunications services have been supplied to the mass market and are
considered a commodity. The trend toward flexible pricing packages is continuing with the most important tariff innovation being prepaid schemes
that sell blocks of airtime in advance of use.[16] Prepaid services are now an
increasing part of the subscriber base for most mobile operators. Their
attractiveness stems from the control they give users over expenditure and
the smaller customer acquisition and billing costs for mobile operators and
also because of reduced scope for fraud and bad debts. To sum, the growth
rates in mobile users reflect trends in the pricing of services related to the
underlying market structure. Entry and more intense competition have led
to the innovation of flexible tariff packages targeted at different categories
of users rather than price cuts.[17] This innovation was mainly responsible for
the first wave of growth and was possible due to the move from analogue to
digital technology that provided additional capacity. In terms of pricing
strategy, a factor that might explain cross-country differences is pricing
structure. While the majority of countries adopted a system whereby the
person initiating the call bears the cost of the call (calling party pays, CPP)
there are important exceptions (Canada and the US) where the receiver also
directly contributes to the cost of a call (receiving party pays, RPP). Under
an RPP system, consumers might be more reluctant to subscribe, keep their
handsets switched off or be less inclined to give away their number.

These considerations help explain why digital mobile services developed
faster in Europe than in the US, the other main reason being the adoption
of incompatible standards. RPP pricing also makes prepaid cards less
favourable to some consumers, since it eliminates most of the appeal of
such schemes – that is, budget control is reduced.[18] Both Canada and the
US have recently reviewed RPP, removing some regulatory barriers to the
introduction of CPP (mainly notification procedures for users and billing
systems). However it is not clear whether CPP will be introduced commer-
cially, as an optional pricing structure, and so far the initial trials have not
proved very successful.[19] Operators in the US have already responded by
launching 'bucket plans' where the customer buys monthly buckets of
minutes on a nationwide network and pays a single rate wherever the call is
placed regardless of where the call terminates. Such plans may reduce the
need to adopt tariffs based on CPP.

Coverage

Mobile telecommunications services are supplied using a grid of cells connected to the fixed network. As areas are covered, or as traffic increases within a given area, the only option an operator has is to invest in additional cells, thus making the cellular technology subject to constant returns to scale.[20] As spectrum is released and licences are attributed to entrants it can be argued that the market will fragment and prices will be brought in line with costs. However, this conclusion should not be taken for granted since consolidation may occur for other reasons. In particular, the major feature of a mobile telecommunications network is its coverage. Calls can be made only if there is a cell covering the area the user is travelling through. Consumers who frequently travel evaluate coverage differently, according to the utility they derive from completed calls. However, even when users differ in the intensity of their preferences, given networks of different size with the same price structure, it is probable that consumers would select the network with wider coverage. In other words network coverage is a parameter similar to a vertical quality difference. This holds as long as customers are mobile enough. Conversely, if a customer is always located in a narrow area covered by competing networks then only the price matters – for these customers, networks are seen as homogeneous goods.

Coverage differentiation has important implications in terms of market structure. In particular, the quality of network coverage implies that natural oligopolies could emerge in the industry.[21] Valletti (1999) shows in a duopoly model that price competition can be relaxed by building networks of differing coverage. In particular, differences in service characteristics are reflected by differences in prices, operators with national coverage are relatively more expensive than operators with smaller coverage, and the nature of price competition is affected by a minimum coverage requirement set by the regulator.[22] This analysis assumes price competition among firms – in practice firms are constrained by capacity limits. The scarcity of the spectrum of radio frequency could be modelled using quantity rather than price competition, and it is well known that quality differentiation is milder the less tough is price competition. In practice, continuous improvements have allowed for a more efficient use of the spectrum, so that while a quantity (Cournot) model would be more appropriate for the past, in the future a price (Bertrand) model should be closer to reality.

Convergence and Bundling

Fixed and mobile telecommunications used to be separate markets. Apart from the different user access technology, the main difference lay in their distinct regulatory frameworks with mobile telecommunications liberalization occurring much earlier. Mobile services can be considered a substitute for fixed services. Digital mobile networks can cater for voice and data and provide many enhanced services available on fixed networks. In practice, however, the extent of substitution is limited by the higher prices and lower quality of mobile calls. Despite this fact, mobile telephony has had a considerable impact on the communications sector's productivity performance. This is because most of the calls to and from a mobile service start and finish primarily on fixed line networks. Accordingly, there is substantial inter-connectivity between alternative components of a network-based infrastructure (Jha and Majumdar, 1999).

Price setting is different in mobile and fixed telephony; for example, domestic calls are generally not distance sensitive with mobile telecommunications (OECD, 2000). Local calls are much more expensive with mobile telephones whereas domestic long-distance mobile calls may be cheaper than calls placed with fixed networks. These differences are bound to become less, as fixed-line operators are rebalancing local call charges to satisfy the regulatory objective of eliminating cross-subsidies and bringing tariffs in line with costs. Moreover, mobile firms have started to introduce differential pricing on domestic calls. For example, home zones are low tariff areas defined in terms of nearness to the home cell. Soon substantial mobile premiums are going to fall and mobile speech quality and data throughput will become comparable to that of the fixed network. These changes will bring about a fusion of fixed line and mobile telephone infrastructure and there will be a blurring of the boundaries between fixed line and mobile operators. Nevertheless, the connecting link between mobile telephones will be a fixed-line link.

In many countries the number of mobile telecommunication subscribers has overtaken the number of fixed telecommunications lines. However, the bulk of the traffic, in terms of minutes, is still with fixed-line telecommunications. This observation is important in the current debate concerning fixed mobile convergence (FMC). A distinction should be made between the way a subscriber obtains access to a service and the way the signal is carried over networks. In terms of having access to a service, mobile and fixed telephony can potentially be substitutes. In this respect, convergence corresponds to market widening manifested by an equalization of functionality, higher cross-price elasticity and approximate pricing between fixed and mobile services. Another distinction should be made between substitution on a call-by-call basis and replacement of the fixed exchange line by mobile telephony. Call-by-call substitution is readily available, however, replacement is currently relevant only for particular market segments – for example, young persons and holiday residences.

When fixed and mobile access have the same bandwidth, mobile services may command a mobility premium both in origination and in reception. This factor introduces a vertical quality differentiation parameter that affects the type of competition between mobile and fixed access operators. Another difference between access technologies arises from their costs. The cost of a fixed-line technology depends mainly on the number of subscribers rather than on their traffic level, while for mobile access technology costs depend on the traffic not on the number of subscribers *per se*. Cremer et al. (1996) argue that under these assumptions it is efficient to employ mobile access technology for consumers with a low willingness-to-pay, since it involves a cost lower than that of the fixed technology for small consumption levels. For moderately high consumption levels the more cost-effective technology is the fixed line. However, for very high consumption levels supply is more efficient with mobile access technology since these customers value mobility most.

FMC involves different considerations when conveyance of data between switching centres is considered. Convergence represents the seamless availability of services, relying both on wireless and wire-line platforms. In this context, mobile and fixed services are better seen as complements – where the networks supply several components that can be combined to supply the final service. The literature on the economics of networks (see Economides, 1996 for a survey) points out that operators have an incentive to integrate network components. This scenario raises important issues concerning standards, licensing, a level playing field among operators and coherency of regulatory approaches. On the one hand, convergence widens markets and introduces further competition, calling for a less intrusive role of regulation (both on mobile and fixed telephony). Conversely, convergence also makes possible the integration of components, increasing the risk of dominance and abuse.[23]

Business propositions are now emerging that offer bundles of fixed and mobile service. FMC is still a floating concept as convergence can be defined at different levels. At the marketing level common use is made of key service elements such as sales force and billing, and at the service level a bundle of fixed and mobile services is provided without linking networks. In this case convergence is perceived by the customer – for example, as a single telephone number – irrespective of the system used and location sensitive pricing. At the network level this entails the complete integration of the fixed and mobile networks, such as single switching. The most appealing feature of FMC to consumers is to have a single handset and number, irrespective of whether the service is provided by a fixed or mobile telephone. This outcome can be achieved by convergence at the service level. For firms the advantage in offering FMC is in increasing the loyalty of consumers through the supply of service bundles.

However, there remain several technological and regulatory hurdles to the supply of FMC services. The cost and uncertainty of technical solutions are high. Moreover, in several countries former monopoly fixed-line operators have been prevented from offering FMC services because of regulatory concerns. The main concern is the possible leverage of market power in fixed-line services to cross-subsidize the more competitive mobile market segment. So in many countries the dominant fixed operator had to divest from mobile operations or at least to proceed to structural or accounting separation. Mobile operators and new fixed network operators are not limited in this way.

As the technical performance of mobile telecommunications, in terms of data transmission, becomes comparable to fixed lines and the pricing becomes more similar, the appeal of FMC should diminish. Mobile operators will be able to offer the same services as fixed-line firms, but with mobility. Some commentators expect that FMC will not become an important market, but it may provide a tool to differentiation services (Jagger, 1998). The incentives and modes for entry into FMC depend on the type of firm. The incumbent fixed operators have the least incentive as they are locked into a dedicated fixed-line capital stock. Entrants have an advantage through designing their network and they can equip to supply FMC from the start. As infrastructure integration is a major hurdle perhaps the greatest chance for this market is firms without dedicated facilities, such as Mobile Virtual Network Operators (Oftel, 1999a).

REGULATION

Regulation of mobile telecommunications services is minimal when compared to that of fixed telephony. Part of the extraordinary success of mobile service growth arguably lies in this lack of intrusive oversight. However, several fundamental regulatory issues, yet to be completely addressed by the mobile industry, remain.

Regulatory Bodies

Regulation of mobile service concerns radio frequency management and technical standards. Because radio waves are not usually affected by national boundaries frequency policy has to be coordinated internationally. Regulation of radio frequency is necessary to avoid interference and because it is scarce. The largest international frequency management and technical standards agency is the ITU. To ensure further harmonization in Europe, issues are dealt with by the European Conference of Postal and

Telecommunications Administrations (CEPT), a body comprised of policy makers and regulators. CEPT handed responsibility for frequency issues to the European Radio Communications Committee (ERC) that has a permanent body – the European Radio Communications Office (ERO). The ETSI is mainly in charge of harmonization and standardization of equipment used. These institutions propose voluntary standards and governments may endorse those standards as mandatory (Bekkers and Smits, 1997). Regulators also have to make decisions concerning the structure of the market by fixing the number of firms and their mode of entry. Post-entry regulation in the operation of mobile networks has tended to be minimal. Mobile firms are typically not affected by universal service and access concerns. In most cases exemption has been granted as mobile telecommunications have been considered a value-added service, and as falling outside the regulatory scope of basic voice service.

Standard Setting

In markets without network effects it is desirable to allow competing technological systems, however, this is less so in markets with network externalities. The presence of strong network externalities typically leads to 'tipping' markets where the winning technology takes the whole market. The theoretical literature does not provide an unambiguous answer to the question of whether the prevailing technology will be the best (for an overview see Katz and Shapiro, 1994). Advocates of government intervention argue that imposing a single standard makes it possible to realize network externalities faster and reduce technological uncertainty among consumers. Free market advocates argue competition will best guarantee better technological systems (possibly a voluntary standard), and reduce the risk of lock-in to an inferior government promoted technology (mandatory standard). A counterargument is that free markets may also lead to lock-in to inferior outcomes, thereby necessitating government intervention to cope with this network externality. That is, one side believes standards generate markets while the other believes that markets generate standards.

Network effects mostly arise from mobile customers only being able to use their handset within areas that support their technological system. Consumers who frequently travel gain from an international standard. Depending on customer mobility, network externalities are local, national and international in scope. In addition to reducing consumer switching costs and creating roaming possibilities, the presence of a single technological system also allows the exploitation of economies of scale in the manufacturing of equipment. Shapiro and Varian (1999) argue that network effects in the cellular mobile industry are strong but not overwhelming.

Licensing

Since the spectrum of radio frequency is limited, governments need an assignment procedure to avoid congestion and interference. Many incumbent fixed telephone operators obtained free mobile service rights as they were seen as an extension of existing services. Subsequently, licences were given after comparative hearings (beauty contests) or lotteries (in the US). Currently, auctions appear the best practice to award spectrum licences. Auctions ensure assets gravitate toward those who value them most highly.[24] For a mobile licence (and associated spectrum) this might be the operator with the greater technical or marketing knowledge, but it might also be the operator with the greatest market power. Consider a situation in which several mobile licences are to be auctioned. The licences probably have highest value in the hands of a single operator, a monopoly service. This outcome is undesirable and can be remedied through auction design – in this case, a rule that an operator can control only one licence.[25] The auction rules can be set to achieve other objectives. For example, in order to bring entrants into a market some licences can be reserved, or they might receive special benefits in the auction process such as having a notional monetary amount added to their bids. Other policy objectives can be built into an auction process by imposing appropriate licence conditions. Should fast network roll-out be desired then a licence could be allocated subject to roll-out conditions – this would presumably reduce firms' willingness to pay, but the government must believe this to be worthwhile otherwise it would not have imposed the condition.

Thus auctions are a flexible means of allocating resources in ways designed to achieve a range of policy objectives. They are also transparent. When auction laws are explicit potential bidders know in advance the basis on which they are competing. This is efficient as it encourages participation. It is also equitable as bidders bid what they want and taxpayers share the revenue. The mode of operation of a beauty contest is quite different. Typically, the government invites applications that are scored according to preset criteria. Licences are allocated to those whom the government believes best meet stated requirements. This procedure has several disadvantages. A subjective element is inevitably present in comparisons, opening the door to favouritism and corruption. Moreover, the government by relying on its own judgement rather than those of commercial interests is implying it has better information about firms' prospects. Even if the jury is perfectly benevolent there is no incentive on the operator's side to tell the truth about its own valuation of licences. Finally, when the administrative charge for a licence is less than its market value forgone auction income accrues to shareholders of the firms. This does not involve

inefficiency but shifts capacity rents between firms, consumers and government.

A cited reason for the use of beauty contests, rather than auctions, is that
auctions typically involve higher licence fees that *ceteris paribus* mean
higher prices for consumers. When the reason for high licence fees is that a
monopoly is being sold, then bids are higher than in a competitive environment because post-auction prices are expected to be higher. But the direction of causation is different. *Ex ante* future prices have an impact on an
operator's willingness to pay for a licence, however, *ex post* prices are not
affected by the licence fee. A firm in deciding what to bid for a licence knows
from the auction rules how many competitors will be licensed and competing to provide service. By forming a conjecture about how the competitive
process will play out, it can estimate what revenue, over and above capital
and operating costs, it will earn. On this basis, it can calculate the maximum
a licence would be worth to it and bid accordingly. A successful firm pays
for the licence either up front in full or through an instalment commitment.
The firm and its competitors consider the licence fee an irrevocable sunk
cost. When deciding how to set prices, the firm rationally only takes account
of its own forward-looking costs, revenues and the likely behaviour of other
firms. Since the licence fee is a sunk cost for all firms, it falls out of the
pricing equation. Hence the size of the licence fee does not affect prices.

Auctions work on the assumption that the objects for sale are easy to
define. In the case of the spectrum, although there are several open questions (transferability, interference and so on), the objects are relatively well
defined. In other circumstances, it could be very difficult to specify all the
characteristics of the objects. In that event, the best procedure could be
direct bargaining between the government and the buyers.

Interconnection

Mobile telecommunications involve a two-way network: calls initiated by a
subscriber of a certain network may be terminated on a different network
and conversely a network will terminate calls originated in other networks.[26] A problem arises from the common practice that the party making
and paying for the call does not choose which operator terminates the call
(CPP). Once a person subscribes to a mobile operator that operator has a
monopoly position over termination services to the subscriber. Clearly, termination services involve an externality that is a potential source of distortion. Subscribing to a network influences the price charged to customers
wanting to call the new subscriber.[27] The termination problem is common
to all network operators. In the context of mobile telephony, the subscriber
base of fixed users is large hence the number of calls potentially terminated

on mobile networks represents an important source of revenue for mobile operators. The presence of such revenues has an important influence on how mobile operators compete.

When the mobile sector is perfectly competitive and mobile operators charge two-part tariffs to customers with identical preferences – for example, a monthly fee and a charge per minute for calls made – then operators compete to attract customers by setting call charges equal to marginal cost. This is a common outcome when firms set two-part prices because operators compete to supply 'value' to their customers and value, given by the sum of the operator's profits and customer surplus, is typically maximized when the price is set equal to marginal cost. The fixed component is used to divide the surplus between the operator and its customers. When the industry is perfectly competitive operators earn zero extra profits. Hence any increase in termination profits, because the termination charge is set above cost, would be passed to mobile subscribers via lower fixed charges. Subscriber fixed charges may even become negative as long as considerable extra profits arise from call termination. This result may explain why handset subsidies are a common feature in many mobile markets.[28] Even with perfect competition for mobile users there is no competition for providing access. This remark suggests that if mobile operators are free to determine termination rates they will set charges that extract all possible surplus from fixed users. In principle some regulatory intervention can be beneficial. This picture also emerges when operators compete less than perfectly. Competition for access revenue induces operators to lower rentals. As long as an increase in the termination charge increases the profit from termination it results in greater competition for cellular customers: firms lower fixed fees to capture market shares and a greater share of termination revenue.[29] In particular, when there is full market participation termination revenue is a perfect substitute for retail revenue, and mobile operators are no worse off with a change in the common termination charges. This situation is counterfactual or mobile operators would not oppose competition authorities' proposals for termination rates more in line with costs. This is probably due to partial market participation: lower fixed fees increase market penetration hence termination revenue is preferred to retail revenue and cellular profits increase as termination charges are raised above costs.

Welfare considerations on termination rates are complicated, since an increase in termination charges both increases fixed-to-mobile calls and decreases mobile fixed fees. As discussed by Armstrong (1997), marginal cost pricing (implying no subsidy for mobile connection) is the correct benchmark when a number of stringent assumptions are satisfied. In particular, the demand of mobile subscribers should be rigid with respect to subscription decisions, there should be no monopoly power exercised by the fixed network and there should be no network or call externality. When mark-ups over termination charges are added by fixed operators they should be counteracted by setting termination charges below cost. Conversely, above-cost charges are beneficial in the presence of network externalities since higher termination revenues can be used to subsidize entry, so raising the equilibrium number of subscribers. As a result, the marginal cost benchmark is appropriate only when there is 100 per cent market participation and no distortion arising from the fixed network. On the contrary, unregulated above-cost termination charges may be 'good' in the initial phases of mobile development since they increase cellular penetration rates.

The potential market failure associated with termination services may be considerably diluted if users care about receiving calls. Implicit in the discussion above is the assumption that users either do not receive or place any value on incoming calls. On the contrary, it is plausible that mobile telephones are purchased to receive calls as well as to make them. Should a mobile user place similar weights on calls made and received then any attempt to set high termination charges would induce a change in network since subscribers would otherwise receive too few calls. This result holds even when the caller and the receiver do not belong to the same 'closed group'. The argument that people wish to receive calls is particularly compelling in the mobile sector where a customer has the ability to be reached at any time and place. Unfortunately there is not enough econometric evidence on calling patterns in mobile telecommunications to indicate the importance of the termination problem.

Another type of interconnection concerns roaming services. When firms are competing for the same market they can relax price competition by building networks with different coverage. In that context national roaming agreements should not happen irrespective of whether agreements are reached in a reciprocal or non-cooperative way. This is because firms would lose their differentiation capability and price too aggressively. Firms should seek roaming with foreign operators since it induces a market expansion effect. International roaming agreements are common in the mobile industry, as long as operators use compatible standards, since there is a double coincidence of wants. This remark is valid so long as operators remain national providers. A recent wave of international mergers and acquisitions, most notably the AirTouch–Vodafone–Mannessman group, may change incentives to provide international roaming. Should global operators face each other in their national markets they may become more reluctant to grant roaming. Denying roaming, or providing it under exceptionally high charges, is a rival cost raising strategy and does not allow effective competition for corporate customers that value coverage highly.

As a result operators may find it necessary to bypass international roaming by either investing in the foreign country, merging with or buying a foreign operator. The roaming problem will increase when 3G operators commence services. Entrants will be disadvantaged when competing with 2G operators that have established customer bases and 2G networks. Since they will be competing within the same voice telephony market it is unlikely there will be voluntary roaming agreements. For these reasons national authorities are having discussions to mandate roaming for a limited period of time to help level the playing field for entrants. In the UK, Oftel (1999b) mandated roaming for seven years. However, entrants are required to roll out networks that cover 20 per cent of the population before roaming is made available to them. The requirements are introduced to commit entrants to substantial investment and avoid pure resale. Roaming charges should be set at 'retail minus', a charge equal to the retail price minus any saving for those elements of the retail service which the roaming operator will substitute for the network operator.[30]

Cost of Regulation

Because of the nature and scarcity of the radio spectrum resource some regulation is indispensable, in particular spectrum allocation in the mobile telecommunications industry. Hausman (1997) argues that delays can be costly. In the US delay was mainly due to regulatory and political indecision as to whether to have a monopoly or duopoly structure. Apart from the social welfare cost, due to the late introduction of a product, US industry lost leadership in mobile telecommunications technology as Japan and the Nordic countries were quicker to grant licences to mobile operators. Another aspect of pre-entry regulation is the setting of standards. Empirical research related to the welfare effects of standard setting in mobile telecommunications is still emerging. Gruber and Verboven (2000b) find that a mandatory standard accelerates the diffusion of mobile telecommunications during the analogue period, lending support to the hypothesis that standardization reduces information costs to consumers and induces more price competition by creating a level playing field for firms. This link was weakened during the digital phase. In the US there was no mandatory national standard and this market generated CDMA technology – the key building block for 3G mobile telecommunications. The cost of standard regulation is therefore the risk of locking into inferior standards.

Post-entry regulation mainly concerns prices. Clearly, moving from a monopoly framework to a less concentrated market structure should provide scope for the relaxation of regulation. However, scepticism exists as to the degree to which a limited number of firms can competitively price.

This argument is based on the large variation in the degree of price regulation that is observed across countries. In the US, price regulation is based at the state level and provides an ideal test bed for the comparative study of the effect of regulatory regimes. Shew (1994) finds prices are generally higher in states with regulated markets. Moreover, rate of return regulation regimes increase prices more than price cap regimes. Also the threat of regulation is the most effective tool for low prices. Parker and Röller (1997) also find that prices are above competitive levels and that mark-ups increase with cross-ownership and multi-market contact.[31] This finding suggests that the public interest requires strong enforcement of mergers, and anti-trust policy may be appropriate, even for geographically distinct markets defined by local licences for radio spectrum. Given that regulation has a positive effect on prices firms may lobby for regulation as it facilitates collusion. Duso (2000) treats the regulatory regime as endogenous and finds that regulation reduces prices. The problem is only that the wrong markets have been regulated – markets with no regulation would have had lower prices with regulation and markets that are regulated will have higher prices than those that would have been observed if there had not been regulation. Selection of the regulatory regime is attributed to the lobbying activity of firms. Overall regulation remains important for this industry and mistakes are costly. Regulatory delays at the pre-entry stage are costly because of relative market inertia and regulatory failure at the post-entry stage exposes customers to the market power of firms.[32]

CONCLUSION

The chapter shows that the development of the mobile telephone industry is conditioned by the availability of radio spectrum. Technological innovation, such as the introduction of cellular technology and subsequent switch from analogue to digital transmission, has helped make the constraint less severe. Further, innovation has helped to extend cellular mobile telecommunications into higher frequency bands of the spectrum and so provided more scope for accommodating new subscribers. On the supply side, relaxing the spectrum constraint has permitted an increase in the number of firms, with beneficial effects on service quality and prices. Very high growth rates have seen the number of mobile telecommunications users exceed the number of fixed lines in many countries. Nevertheless, diffusion of mobile telephony is uneven. The empirical diffusion literature finds that typically digital technology and entry of firms accelerate diffusion, and that differences in diffusion persist.

The introduction of mobile telecommunications has strongly impacted on the telecommunications industry and was instrumental in many countries in breaking up the monopoly structure of the sector. Some of the most striking changes occurred in pricing, where a full range of pricing patterns emerged to segment the market as it became larger. Other means of non-price competition concern service coverage and bundling with other services such as fixed line telecommunications services. Although prices have been falling, empirical analysis indicates that market structure is far from competitive. Market conduct studies suggest firms tend to practise non-competitive pricing sustained by tacit collusion. Regulatory tasks can be distinguished between pre- and post-entry. Pre-entry regulation concerns the setting of technical standards and entry conditions. Setting standards is perceived as socially beneficial because it induces firms to place more emphasis on price competition and so reduces consumer search costs. However, setting standards may lock the market into sub-optimal standards and hinder the development of technology. The emergence of the CDMA technology may provide an example. There is agreement that social welfare increases with the number of firms. However, it is not clear that increasing the number of firms is the most efficient way to achieve lower prices. This goal might be better achieved through post-entry regulation. In any case, it may be difficult to abolish post-entry regulation because of the persistence of the scarcity of radio spectrum. Firms are therefore likely to retain considerable market power. The mobile telephony industry has moved toward replacing regulation by competition. The reason for this is that the benefit of regulation has to be traded off with that of regulatory failure. Empirical evidence for US duopoly markets suggests regulation is often affected by lobbying and leads to higher prices. Overall the cost of regulation seems high. However, more empirical analysis is required on the effects of regulation using evidence from an industry with more than two firms.

Although the rapidly increasing literature on the economics of mobile telecommunications has given several new insights, there are many interesting research questions that still have to be addressed and which are also of broader interest. The fact that technological cycles are very rapid, and take much less time to unfold than in other industries, makes this industry particularly interesting in testing propositions taken from economic theory. Empirical research is facilitated by the fact that it is well documented and that there is a wide range of neat national settings that allow for rewarding cross-country analysis. This is the more interesting as the industry experiments with new regulatory instruments, such as auctions, that have important implications also for other industries.

NOTES

The opinions expressed are of the authors and need not necessarily reflect those of their respective affiliations.

1. Early mobile systems had limited capacity since they made use of the spectrum in an inefficient manner. Systems were based on the same principles as radio or television broadcasting. They used high power transmitters located in base stations to gain maximum coverage. Transmitters operated at very low frequency levels of around 150 MHz. At such low frequencies, signals travel far and base stations have a large coverage area with a radius of up to 80 km. This means that only a few base stations are required, however, the few available channels that support conversations are locked up and can only serve a small number of users. These principles are valid in general, not just in the early days.

2. Calhoun (1988) reports that the early mobile communications industry struggled to gain any frequencies in the 100 MHz band during the 1950s and 1960s. Analogue cellular technology had to wait until the beginning of the 1980s for spectrum in the 400 MHz band to be granted. Hausman (1997) estimates the economic cost of this regulatory delay at 30 to 50 billion United States dollars (USD) per annum for the US. The frequencies for digital cellular service such as GSM became available much quicker in the 900 MHz band at the end of 1980s and in the 1800 MHz band around mid-1990s (Bekkers and Smits, 1997). For the 3G mobile systems the huge economic value of mobile service and prospective licence fees induced policy makers to provide spectrum in the 1900 MHz range (Gruber and Hoenicke, 1999).

3. Coverage of a large geographic area is obtained by dividing it into cells. Many cells with many base stations are required to obtain full coverage of a large geographic area. This implies considerable investment. A crucial advantage is that the channels to support simultaneous conversations are only locked up over a limited cell area; that is, the frequency channels can be reused to support telephone conversations in other cells. This can greatly increase capacity. To avoid interference, adjacent cells have to operate on different sets of frequencies while frequencies can only be reused in nonadjacent cells. As a mobile user moves from cell to cell the continuity of the conversation must be ensured. This requires a change in frequencies of the specific radio channels used for transmission from each base station (handover). When an existing cell has reached its capacity it can be further subdivided into additional cells (cell splitting). Cell splitting increases the scope for frequency reuse (see Garg and Wilkes, 1996; Rappaport, 1996).

4. For an illuminating description of the emergence of the NMT standard see Mölleryd (1997).

5. For a description of national markets see Schenk et al. (1995) and Garrard (1998).

6. For a background to the creation of the GSM standard see Redl et al. (1995).

7. On the competition among cellular systems in the US see Olmsted Teisberg (1992) and Shapiro and Varian (1999).

8. CDMA is technically superior to TDMA for data transmission. CDMA sends coded signals on a broad band of frequencies and uses handsets that only listen to its own code.

9. A precursor for Internet access by mobile telephone is i-Mode, a service launched in 1999 by the Japanese operator NTT DoCoMo, which gained more than 10 million users in Japan by mid-2000. The i-Mode mobile data services enable users to do telephone banking, make airline reservations, conduct stock transactions, send and receive e-mail and have access to the Internet. At the moment WAP and i-Mode are based on incompatible protocols but there are attempts to make them compatible to avoid a standards war.

10. Capacity constraints imply that only a limited number of calls can be handled simultaneously in a cell. When the density of cells is low cellular users face the probability of having their conversation cut off, especially during peak hours.

11. Salop and Stiglitz (1977) examine the case of a single monopolist selling through several outlets. When consumers know only the distribution of prices charged at various outlets

but not precisely which stores charge which prices, there exist multiple equilibria involving price dispersion. For some parameter configurations there are exactly two prices charged in equilibrium: low-price stores compete for the informed customers and high-price stores exploit the uninformed customers.

12. The shape depends on the parameters of population of potential adopters, timing of adoption and speed of adoption.

13. For example, Valletti and Cave (1998) describe the development of competition for the UK.

14. Parker and Röller (1997) consider a panel of 305 US metropolitan markets during the early stages of network growth when most were monopoly and duopoly markets (see also Ruiz, 1995). Cross-sectional variation provides useful insights into the nature of competition in mobile telephony. They find that prices were considerably above competitive duopoly levels (prices included a 35 per cent mark-up over marginal costs). They also find that collusive behaviour was explained by structural variables, in particular, operators were more likely to collude if they faced each other in several markets (multi-market contacts) since deviations from a collusive strategy would lead to a chain effect across markets. Collusion was also more likely when there was cross-ownership, that is, firms competing in a market were also partners in some other market. In the US fewer restrictions on co-ownership have been placed on independent operators rather than RBOCs. This might explain the finding that independent operators were systematically colluding more than RBOCs. These results suggest something similar may happen once global mobile operators emerge.

15. The problem of designing contracts that can screen among consumers with multidimensional preferences in a competitive environment is an ongoing area of research. See Armstrong and Vickers (1999) and Rochet and Stole (2000).

16. Prepaid cards were introduced in Germany and Switzerland in 1995 but were not rechargeable. Their initial commercial success was due to the marketing strategy of Telecom Italia Mobile (TIM) that adopted rechargeable cards in 1996. In June 1999, 80 per cent of TIM users and most new users were prepaid. On the other hand, only 6 per cent of subscribers use prepaid plans in the US (FCC, 2000).

17. Mobile telephony is still quite an expensive service. Fixed network operators have responded to falling mobile telecommunications service prices by cutting fixed-line call prices. For instance, in Scandinavia, where mobile telephony is well established, mobile services are on average five times more expensive than fixed telecommunications services based on per minute prices (OECD, 2000).

18. Alternatively, an RPP system is transparent and puts more pressure on operators to cut charges for call termination, since both incoming and outgoing calls are paid by the person who chooses the mobile network operator (see Doyle and Smith, 1998).

19. In Mexico a regulatory decision introduced CPP from May 1999, having adopted RPP until then. Although only limited data is available this innovation of CPP has coincided with record growth.

20. For the US, McKenzie and Small (1997) find some diseconomies of scale and constant return to scale at best. Foreman and Beauvais (1999), employing a richer set of data, find mild scale economies. The results relate to technology as marketing and administration costs are not considered.

21. Expenditure on cells is endogenous and by making such outlays an operator can enhance the demand for its product. Anticipating operators might engage in an escalation of investment, leading to higher sunk costs at equilibrium. As market size increases, the investment escalation in earlier stages will raise the equilibrium level of sunk costs incurred by incumbent firms, in line with increases in the size of the market, thus offsetting the tendency toward fragmentation. This is confirmed by the emergence of operators with nationwide coverage in the US (AT&T Wireless, Sprint PCS and Verizon, Voicestream) or coverage that extends beyond national boundaries (Vodafone Airtouch).

22. The crucial role played by coverage is confirmed in the UK. The incumbent operators BT Cellnet and Vodafone reached full population coverage of both their digital and

analogue networks quickly. Cheaper PCN operators (Orange and One2One) followed with a strategy of differentiation. In particular, One2One covered only 40 per cent of the population five years after having been awarded a licence (Valletti and Cave, 1998). More recently, PCN operators had to revise their plans for the roll-out of their networks as firms succeeded in matching their coverage, so resulting in decreased product differentiation and intensified price competition. A similar pattern emerged in Germany where firms initially managed to avoid price competition by product differentiation (Nattermann, 1999). It was the reduction in the scope for differentiation rather than the increased number of firms that led to falling prices.

23. Reiffen et al. (2000) find the degree of integration between the (often regulated) fixed-line operators and unregulated mobile operators has a significant impact on prices, quantity and quality of cellular telephone services. Integrated firms are more efficient in the provision of service but tend to discriminate against mobile rivals needing interconnection with the fixed network.

24. See Cramton (1997), McAfee and McMillan (1996) and McMillan (1994). Also see Klemperer (2000) for a recent analysis of the 3G auctions in Europe. In a sense the focus on licence assignment misses the point as licences are assigned to particular operators within the context of a prior allocation of spectrum across major types of usage. When this spectrum allocation is made sub-optimally the spectrum assigned for some specific purposes is unnecessarily scarce.

25. The observation answers one of the criticisms made about auctions – that they are adopted to maximize revenue for the government. When the government really wants to maximize revenue it should auction monopoly rights and not five licences for competing services as happened in the UK for UMTS.

26. Interconnection charges for terminating mobile calls on the fixed network are not considered here since they are typically regulated. Note that network interconnection in principle eliminates network externalities generated by the desire of users to call other users. However, mobile operators have started to offer packages with different prices according to whether a call is destined to users on the same network. Tariff-mediated network externalities are increasingly important as the mobile subscriber base becomes comparable to the fixed user base.

27. Another source of distortion in the termination of fixed-to-mobile calls is consumer ignorance (Gans and King, 2000). In its inquiry into mobile termination rates the UK Monopolies and Mergers Commission found fixed-line users had little knowledge of the mobile network they were calling and of call price (MMC, 1998). When fixed-line users make decisions on estimated prices based on mobile market shares then the link between a specific termination charge set by a network and the number of calls terminated on that network is broken. When a mobile network raises its termination charge it obtains the full benefit and shares with other mobile networks the reduction in the number of calls received. Consequently networks have an incentive to set high termination rates. This problem is exacerbated by the adoption of mobile number portability as there can be no correspondence between a telephone number and the current subscribers network. Carrier identification should be promoted to make termination services more competitive.

28. Handset subsidies and subscription discounts are also related to the presence of switching costs. Switching cost are non-recoverable costs that are incurred when a consumer changes the supplier (Klemperer, 1995). With consumer switching costs a firm is willing to serve a larger customer base than found in traditional models because such behaviour enlarges its captive segment of the market. After securing a customer base a firm restricts output to exploit its limited monopoly obtained from switching costs over its customers. Thus the practice of subsidizing terminals is a plausible strategic investment in customer base establishment when the new customer produces cash flow for the operator. For mobile communications additional switching costs are incurred when customers are connected to an operator for a contract period (usually 12 months for contracts other than prepaid cards), when the telephone may not work on another network (hardware lock), and stationary costs in the absence of number portability. The presence of switching costs makes it important to understand the vertical links between operators and service

providers (Valletti, 2000). Discounts are usually offered by service providers as a natural response to network operator incentive contracts. Bonuses give incentives to service providers, or to their associated dealers, to attract customers by offering discounts on handset prices and the subscription fee.

29. This argument is relevant when charges are set cooperatively. When they are set non-cooperatively there would be an escalation game in which termination charges are set above marginal cost (Wright, 2000). Operators prefer to set termination charges higher than their rival to receive higher termination revenues, in order to subsidize lower fixed charges and increase market share without sacrificing per customer profit. Unless escalation is stopped, for example by the threat of regulation, it will continue until demand from fixed users is eliminated.

30. Clearly, 'retail minus' is a version of ECPR (Efficient Component Pricing Rule). Theory suggests ECPR is appropriate when retail tariffs are controlled, which is not the case for mobile communications. Also notice the potential conflict between the regimes typically proposed for mobile and for fixed telephony. Indirect access on mobile networks is based on the opportunity cost of the incumbent, while indirect access to an incumbent's fixed network is typically based on cost-plus using long-run incremental cost principles.

31. Although multi-market contacts may theoretically enhance firms' ability to tacitly collude, firms still need to develop a means to communicate and coordinate their actions. It is particularly important in practical antitrust cases to understand how firms coordinate their prices. Busse (2000) shows that firms in the US, during the duopoly period, used price schedules as their strategic instruments to coordinate markets. In particular, identical price schedules set by one operator across different markets could help operator efforts to tacitly collude. Also price matching, where different firms set the same price within the market, does increase average price. However, price matching is not associated with multi-market contact.

32. Another distortion arises when taxes are levied on mobile service use. Hausman (2000) finds that due to the relatively elastic demand for mobile telephony in the US every USD1 raised in tax imposes an efficiency loss of USD0.50.

REFERENCES

Armstrong, M. (1997), 'A simple model of competition in mobile telephony', mimeo, London Business School.

Armstrong, M. and Vickers, J. (1999), 'Competitive price discrimination', mimeo, Oxford University.

Bekkers, R. and Smits, J. (1997), *Mobile Telecommunications: Standards, Regulation and Applications*, Artech House, Norwood.

Busse, M.R. (2000), 'Multimarket contact and price coordination in the cellular telephone industry', *Journal of Economics and Management Strategy*, 9(3), 287–320.

Calhoun, G. (1988), *Digital Cellular Radio*, Artech House, Norwood.

Cramton, P. (1997), 'The FCC spectrum auction: An early assessment', *Journal of Economics and Management Strategy*, 6(3), 431–95.

Cremer, H., Ivaldi, M. and Turpin, E. (1996), 'Competition in access technologies', DT 60, IDEI, Université de Toulouse.

Doyle, C. and Smith, J.C. (1998), 'Market structure in mobile telecoms: Qualified indirect access and the receiver pays principle', *Information Economics and Policy*, 10(4), 471–88.

Duso, T. (2000), 'Who decides to regulate? Lobbying activity in the US cellular industry', WZB Discussion Paper FS IV 00-05.

Economides, N. (1996), 'The economics of networks', *International Journal of Industrial Organization*, 14(6), 673–99.

European Commission (1997), 'On the further development of mobile and wireless communications', COM(97) 217 final, Brussels.

FCC (2000), Annual report and analysis of competitive market conditions with respect to commercial mobile services – Fifth report FCC 00-289', FCC, Washington, DC.

Foreman, R.D. and Beauvais, E. (1999), 'Scale economies in cellular telephony: Size matters', *Journal of Regulatory Economics*, 16, 297–306.

Funk, J.L. (1998), 'Competition between regional standards and the success and failure of firms in the worldwide mobile communications market', *Telecommunications Policy*, 22(4–5), 419–41.

Gans, J.S. and King, S.P. (2000), 'Mobile network competition, customer ignorance, and fixed-to-mobile call prices', *Information Economics and Policy*, 12(4), 301–27.

Garg, V.K. and Wilkes, J.E. (1996), *Wireless and Personal Communications Systems*, Prentice Hall, Upper Saddle River.

Garrard, G.A. (1998), *Cellular Communications: Worldwide Market Developments*, Artech House, Norwood.

Gruber, H. (2001), 'Competition and innovation: The diffusion of mobile telecommunications in Central and Eastern Europe', *Information Economics and Policy*, 13(1), 19–34.

Gruber, H. and Hoenicke, M. (1999), 'The road toward third generation mobile telecommunications', *Info*, 1(3), 213–24.

Gruber, H. and Hoenicke, M. (2000), 'Third generation mobile: What are the challenges ahead?', *Communications and Strategies*, 38, 159–73.

Gruber, H. and Verboven, F. (2000a), 'The diffusion of mobile telecommunications services in the European Union', *European Economic Review*, forthcoming.

Gruber, H. and Verboven, F. (2000b), 'The evolution of markets under entry and standards regulation. The case of global mobile telecommunications', *International Journal of Industrial Organization*, 19(7), 1189–1212.

Hausman, J.A. (1997), 'Valuing the effect of regulation on new services in telecommunications', *Brooking Papers on Economic Activity. Microeconomics*.

Hausman, J.A., (2000), 'Efficiency effects on the US economy from wireless taxation', *National Tax Journal*, September, 733–42.

ITC (1993), 'Global competitiveness of U.S. advanced-technology industries: Cellular communications', Publication 2646. ITC, Washington, DC.

ITU (1999), 'World telecommunication development report 1999: Mobile cellular', ITU, Geneva.

Jagger, H. (1998), 'Fixed/mobile convergence: Commercial success strategies', Ernst and Young.

Jha, R. and Majumdar, S.K. (1999), 'A matter of connections: OECD telecommunications sector productivity and the role of cellular technology diffusion', *Information Economics and Policy*, 11(3), 243–69.

Kargman, H. (1978), 'Land mobile communications: The historical roots', in Bowers, R., Lee, A.M. and Hershey, C. (eds), *Communications for a Mobile Society*, Sage Publications, Beverly Hills.

Katz, M.L. and Shapiro, C. (1994), 'Systems competition and network effects', *Journal of Economic Perspectives*, 8(2), 93–115.

Klemperer, P. (1995), 'Competition when consumers have switching costs: An overview with applications to industrial organization, macroeconomics, and international trade', *Review of Economic Studies*, 62(4), 515–39.

Klemperer, P. (2000), 'What really matters in auction design', mimeo, Oxford University.

Laffont, J.-J. and Tirole, J. (2000), *Competition in Telecommunications*, MIT Press, Cambridge, MA.

Levin, H.J. (1971), *The Invisible Resource: Use and Regulation of the Radio Spectrum*, Johns Hopkins Press, Baltimore.

McAfee, R.P. and McMillan, J. (1996), 'Analyzing the airwaves auction', *Journal of Economic Perspectives*, 10, 159–76.

McKenzie, D.J. and Small, J.P. (1997), 'Econometric cost structure for cellular telephony in the United States', *Journal of Regulatory Economics*, 12, 147–57.

McMillan, J. (1994), 'Selling spectrum rights', *Journal of Economic Perspectives*, 8, 145–62.

MMC (1998), 'Cellnet and Vodafone', Monopolies and Mergers Commission, London.

Mölleryd, B.G. (1997), 'The building of a world industry: The impact of entrepreneurship on Swedish mobile telephony', Teldok, 28e, Stockholm.

Nattermann, P.M. (1999), 'Estimating firm conduct: The German cellular market', doctoral thesis, Georgetown University.

OECD (2000), 'Cellular mobile pricing structures and trends', DSTI/ICCP/TISP(99)11/final, OECD, Paris.

Oftel (1999a), 'Statement on mobile virtual network operators', Office of Telecommunications, London.

Oftel (1999b), 'Statement on National Roaming', Office of Telecommunications, London.

Olmsted Teisberg, E. (1992), 'Technology choice in digital cellular phone switches', Working Paper 92-062, Harvard Business School.

Parker, P.M. and Röller, L.-H. (1997), 'Collusive conduct in duopolies: Multimarket contact and cross-ownership in the mobile telephone industry', *RAND Journal of Economics*, 28(2), 304–22.

Rappaport, T.S. (1996), *Wireless Communications: Principles and Practice*, Prentice Hall, Upper Saddle River.

Redl, S.M., Weber, M.K. and Oliphant, M.W. (1995), *An Introduction to GSM*, Artech House, Norwood.

Regli, B.J.W. (1997), *Wireless. Strategically Liberalising the Telecommunications Market*, Lawrence Erlbaum Associates, Mahwah.

Reiffen, D., Schumann, L. and Ward, M.R. (2000), 'Discriminatory dealing with downstream competitors: Evidence from the cellular industry', *Journal of Industrial Economics*, 48(3), 253–86.

Rochet, J.-C. and Stole, L. (2000), 'The economics of multidimensional screening', mimeo, University of Chicago.

Ruiz, L.K. (1995), 'Pricing strategies and regulatory effects in the U.S. cellular telecommunications duopolies', in Brock, G.W. (ed.) *Towards a Competitive Telecommunications Industry*, Lawrence Erlbaum Associates, Mahwah.

Salop, S. and Stiglitz, J. (1977), 'Bargains and ripoffs: A model of monopolistically competitive price dispersion', *Review of Economic Studies*, 44, 493–510.

Schenk, K.E., Müller, J. and Schnöring, T. (eds) (1995), *Mobile Telecommunications: Emerging European Markets*, Artech House, Norwood.

Shapiro, C. and Varian, H.R. (1999), *Information Rules: A Strategic Guide to the Network Economy*, Harvard Business School Press, Boston.

Shew, W.B. (1994), 'Regulation, competition, and prices in the US cellular telephone industry', mimeo, American Enterprise Institute.

Sutton, J. (1998), *Technology and Market Structure. Theory and History*, MIT Press, Cambridge, MA.

Valletti, T.M. (1999), 'A model of competition in mobile communications', *Information Economics and Policy*, 11(1), 61–72.

Valletti, T.M. (2000), 'Switching costs in vertically related markets', *Review of Industrial Organization*, 17(4), 395–409.

Valletti, T.M. and Cave, M. (1998), 'Competition in UK mobile communications', *Telecommunications Policy*, 22(2), 109–31.

Webb, W. (1998), *Understanding Cellular Radio*, Artech House, Norwood.

Wright, J. (2000), 'Competition and termination in cellular networks', mimeo, University of Auckland.

9. Satellite communications services

Joseph N. Pelton

INTRODUCTION

At the beginning of the twenty-first century the primary transmission media include satellites, coaxial and fiber optic cable, and terrestrial wireless technology. Satellite technology has exhibited substantial growth in adoption since its commercial inception in 1965 with satellite transmission capability in orbit increased one hundred thousand fold. Current technology is nearly a thousand times more cost effective when measured in terms of throughput times satellite life. Nevertheless the satellite industry exists in the shadow of extremely high throughput fiber optic systems that can deploy systems capable of operating at speeds in the terabits per second. Such fiber optic systems can transmit megabytes of data for under 0.01 United States dollars (USD) and may soon send gigabytes of data for under this price. Certainly the price of transmitting data globally in the twenty-first century is far less than that of printing it out on hard copy. In this environment satellite service providers will compete for specific parts of this nearly USD1 trillion global telecommunications market. Furthermore, while the satellite market will continue to grow in size in conventional fixed satellite services it will exhibit its greatest growth in large-scale broadcasting – that is, DBB/DTH services – in both national and international markets, digital video broadcast and other multi-node networks, and rural and remote services. However, despite this projected bright future a question remains with regard to land mobile satellite services, where market failures, especially the bankruptcy of the Iridium low earth orbit (LEO) systems and mounting problems with Globalstar, suggest such forecasts are optimistic.

A unique feature of satellite systems is that they have the capability to provide emergency warning and recovery operations. This aspect of the satellite industry for emergency communications was recognized in Japan subsequent to the great Kobe earthquake. In many natural disasters such as earthquakes, volcanic eruptions, hurricanes, fires and tsunamis terrestrial cable networks can be destroyed or disrupted due to flooding or rupture. As the cost of disaster related damages continues to rise into the tens of billions of USD annually, the criticality of wireless and satellite communications services in surviving disasters is becoming apparent. Also in the case of some broadcast, rural and remote, navigational, military and mobile services satellites offer capabilities that are not available via other media. This has served to make the market structure for some of these services inelastic. This condition will become less true as terrestrial wireless services, broadband wireless services, high altitude platform systems, or stratospheric platforms and a greater diversity of competitive satellite systems evolve. Currently several satellite services can be provided at premium (value-added rates) because of the unique capability of space communications systems. Within the current decade the large-scale deployment of fiber optic, broadband terrestrial wireless, high altitude platform and multi-purpose satellite systems will see this become less true.

Another factor complicating this analysis is that the information systems, communications and entertainment industries are currently in the midst of tumultuous change amplified by their convergence into a super market. This merging is referred to as the new ICE (information, communication and entertainment services) age. This emerging world of multimedia and merged digital markets is expressed in terms of rapid technological innovation driven by digital systems, convergence of all forms of information markets, mergers of large corporations across international boundaries and significant new patterns of regulatory reform and liberalization of markets. This digital information mega-market can also be characterized as the merger of communications, computers, consumer electronics, cable television and content – that is, movies, newspapers, software and publishing – markets. In a digital age where information is digitally stored, processed and distributed, old market distinctions become obsolete. The range of converging markets driven by digital technology is shown in Figure 9.1.

This chapter analyzes how the world of satellites fits into the overall world of telecommunications, information and entertainment systems from the perspective of technology, regulatory shifts and changing economic trends. The world of satellite communications is driven rapidly by market demand, developments in space and terrestrial communications technology, spectrum allocation, and service innovation. Communications satellites, because of their unique capabilities will most likely play a greater role in some parts of these vast new markets than others. Satellites will, for instance, represent a very significant part of the direct-to-home entertainment, IP multicasting and mobile information service market and less so in the overall broadband communications market, at least during the next five years. In terms of total turnover volume, however, communications satellites may still play a substantial role. This is because satellites are often a

Tele-power service providers		Manufacturers/suppliers
Alternative network providers/teleports		
Cable TV providers		
Electronic information providers and book publishers		Computer and robotic manufacturers
Movie, TV and cable TV programmers		Electronic appliances (e.g. TV) display manufacturers
Inter-exchange carrier – Local telephone providers	SINGLE MARKET	Optical electronic component/ solid state manufacturers
Local exchange carrier – Local telephone providers		Telecom switch manufacturers
Magazine and newspaper publishers		Telecom transmission manufacturers, wire and wireless
Smart energy providers, software and artificial intelligence developers		Smart homes, buildings and town providers
Tele-education and tele-health providers		
TV broadcasters		

Figure 9.1   Market convergence in the digital age

high value telecommunications service because of their mobility, surviv-ability, universality and instantaneous ability to be installed at virtually any location. However, the ability to offer unique services, and to benefit from demand that is at least partially inelastic, will be reduced over time. For example, it is apparent that charges for terrestrial wireless telecommunica-tions services are increasingly moving toward parity with terrestrial wire line telecommunications services. Value-added pricing that allowed terres-trial mobile cellular and PCS services to be highly profitable from the mid-1980s to mid-1990s is being reduced because highly competitive offerings expand consumer choice. Choice is no longer restricted to having or not having the service, but rather to having the service from alternative and highly competitive providers; that is, digital cellular systems, PCS and enhanced mobile service radio.

CHANGING MARKET STRUCTURES

An important trend in the age of digital convergence is the merging of satellite, fiber, and wireless technology within a single global service provider. An example is emerging within integrated global information carriers such as Lockheed Martin Global Telecommunications and Alcatel with access to all types of terrestrial and space communications media. This tendency will lead to a global market for satellites, make fiber and wireless more elastic and commoditize telecommunications. In 1994 the Negroponte Flip forecast that all broadband services would go on to broadband fiber optic systems and that most narrow band services would go on to mobile wireless systems (see Figure 9.2). This technology-driven forecast was based on a perceived shortage of radio wave spectrum. The prediction was essentially a technological imperative and suggested that a lack of additional radio frequency spectrum to support satellite and wireless growth would force the migration of wireless and satellite services for their capability to provide mobile – albeit narrow band mobile. However, technology, driven by market demand for mobile services, found solutions to support broadband wireless and satellite systems growth, as well as broadcasting services via MPEG 2 technology and interactive multimedia. Satellite service sectors have grown by a factor of ten during the past decade, and market demand and technological innovations promise more growth. Instead a Pelton Merge that suggests a mix of media responding to consumer demand is becoming evident (see Figure 9.3). Increased demand for more mobile and

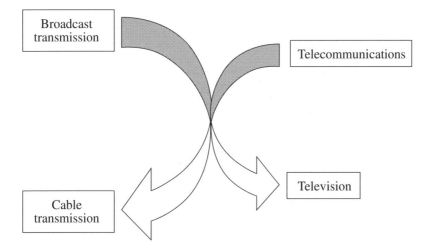

Figure 9.2   Negroponte Flip

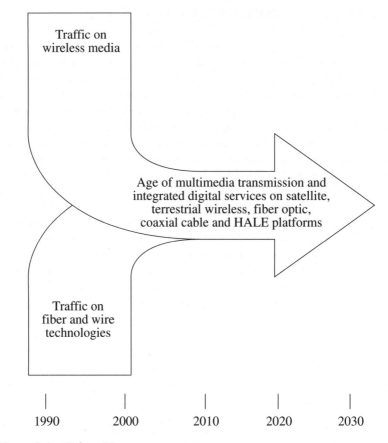

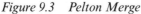

Figure 9.3   Pelton Merge

digital broadcasting, Internet-related IP multicasting services, and a desire
to achieve leapfrogging in a number of developing countries has fueled
much wireless and satellite innovation. These developments have stimu-
lated a growth in wireless and satellite services. In fact, global wireless and
satellite revenue growth, on a percentage basis, have far outstripped the
increase in fiber network revenues.

The emerging liberalized regulatory framework within OECD member
countries, and the expansion of the WTO member countries accepting
open trade in telecommunication services, is more accommodating to a
more competitive global market. This change benefits the growth of wireless and satellite systems that can more easily bypass traditional wire and
cable-based telecommunications systems. In this more liberal environment
satellite service providers more directly interconnect with end users.

The global evolution of satellite communications is highlighted by new satellite innovations that reduce service costs and link end users more closely. Market structures have moved rapidly away from monopoly systems. Entities like INTELSAT, INMARSAT and EUTELSAT have become privatized and open to competition. Perhaps most significantly the economies of the Internet that derive from open networking and IP-based technology have served to bring the computer and telecommunications markets into open competition with one another in the satellite world, just as has occurred in the terrestrial networking world. During the period 1998 to 2000, the revenues for satellite-based IP multicasting (using digital video broadcast technology) grew from USD100 million to USD1 billion. In summary, satellites have benefited from convergence largely created by digital technology. New satellite systems are able to bypass traditional telecommunications carriers and thus have begun to provide entertainment and mobile communications services directly to mass consumer markets. The next wave of satellite innovation will come when advanced, extremely high frequency satellite networks are able to provide broadband multimedia services to the end users and integrate entertainment, mobile communications, data and interactive telecommunications services via very compact and mobile transceivers.

Currently satellites can provide broadcast entertainment services to the household or business at very low cost and bill consumers directly without intervention by a conventional telecommunications carrier. The ability of mobile satellite systems to provide broadband information services is just beginning. It is the hope of new communications satellite providers that offer medium to broadband multimedia services to capture a part of the new multimedia information market. These include new systems such as Hughes' Spaceway, Lockheed Martin's Astrolink, Loral's Cyber Star, Sky bridge and Teledesic, but it will also include the reinvigorated INTELSAT and the more conventional PanAmSat, and GE Americom systems. All these systems, operating in the Ku and Ka bands, will be able to provide global broadband services to meet a variety of consumer needs and in doing so provide direct service to end users. This environment is affected by technology (in space and on the ground), regulatory reform, international trade regulations, global standards, approaches to protecting copyright and intellectual property, pricing and billing arrangements, and other institutional and market shifts. With telecommunications satellites moving toward mainstream global services and able to offer services directly, service will be priced at ever more competitive levels and its value-added price premiums will erode. Satellite service providers will be able to be far more innovative in their pricing and billing arrangements. Another market feature is the ability to provide communications satellite services on demand so that a consumer can shift from data to voice to facsimile to videoconference to broadband video and back to voice on demand. Other changes are the shift from satellite services from being highly limited and rigidly priced common carrier-based services to information products that can be versioned as in the information industry. In the world of versioning different consumers can pay for updated and enhanced services. Internet and IP providers via satellite such as I-Beam, Sky Cache and Edgix are pioneers in the field. Finally, as fiber and satellite systems are increasingly integrated the consumer will be paying for information services, and it will be irrelevant whether the bits come via a wireless, fiber or satellite facility.

COMMUNICATIONS SATELLITE TECHNOLOGY

Rapid advance in fiber optic technology has overshadowed parallel developments in satellite communications systems. Advances in satellite communications systems have come from the development of advanced high gain antenna systems, advanced solar and battery power systems, and digital communications processing systems that are capable of processing and regenerating signals at very rapid speeds. These combined developments have allowed geosynchronous communication satellites to evolve higher capacities as ground systems have shrunk in size and cost. Equally important is that these larger, more powerful and broader spectrum band geosynchronous satellites allow technology inversion to occur. That is, more powerful satellites have allowed the user terminals to shrink in size and cost. Further, demand for mobile satellite services and very small user terminals, combined with the lack of available spectrum to support mobile satellite services and concerns about latency, served to support the use of LEOs. Satellites in low earth orbit are 40 times closer to the Earth surface than geosynchronous systems and offer several advantages. Satellites with smaller and lower gain antennas achieve more tightly focused beams and create a higher flux density at the Earth's surface, and facilitate the creation of many cell-like beams which afford an opportunity to reuse frequencies. In short, the LEO satellites' antenna systems resemble the cellular reuse format found in terrestrial cellular systems. Offsetting disadvantages include the deployment of many more satellites with more launches, and the requirement for more exacting onboard switching systems. The associated higher costs of building and launching the systems, and their lower lifetime, are serious obstacles to their commercial viability. Low and medium earth global satellite systems, such as Iridium and ICO are bankrupt under the Chapter 11 provisions of US law. The Globalstar mobile satellite system is currently struggling for survival and ICO is hoping that capital and market knowledge from billionaire Craig McCaw can redefine this system into a broader band Internet service that can succeed in the global market place.

Advances in LEO satellite technology have seen the evolution of GEO satellite systems with very high power levels and large aperture antennas that offer competitive services for mobile users – for example, Thuraya (that provides services to the Arab world), ACeS (that provides services to the Asian continent) and Agrani (that will service the African and Asia continents). The Telesat F-2 satellite scheduled for launch in 2002 will have 13.5 KWs of power. Emerging trends include exploitation of higher and broader band frequencies, advanced antenna technology, a variety of new orbits, and the increased integration of satellites with terrestrial wireless and fiber optic systems. Improvements in digital compression software allow substantially more bits of information to be transmitted through smaller amounts of spectrum. The current norm for satellite communications is approximately 1 bit per hertz of available frequency. Within a decade modems especially adapted for broadband wireless communications may be able to support 8 bits per hertz.

Technology is no longer a process of innovation in the laboratory that is then patented and implemented. Today the standards making process and the regulation of how systems are designed and operated have a huge impact on how soon and even which new technologies are utilized. Satellite communications are often constrained by standards that are set for terrestrial networks and information systems operation. Large challenges that satellite systems face in this arena include:

(a)   Standards, operating systems and software to allow satellite systems to achieve maximum compatibility with TCP/IP operation.
(b)   Standards, operating systems and software to allow satellite systems to have maximum compatibility with Asynchronous Transfer Mode (ATM) operation.
(c)   Air Interface standards to allow maximum compatibility between satellites and broadband wireless and mobile terrestrial wireless systems.
(d)   Error control systems and protocols that allow satellites maximum seamless connectivity with terrestrial systems (with particular attention to transmission latency, bit error rate comparability and overhead reduction).

As digital technology expands ICE age convergence and as consumer demands accelerate trends toward the Pelton Merge, the pace of technological innovation in the satellite field and of standards development will increase. That is satellite technology and standards must evolve more quickly to stay relevant. As satellite systems are constrained by available radio spectrum, research will be needed in this area. This increases the likelihood that high altitude platform systems (HAPS) may evolve to work in tandem with satellites. These new stratospheric platforms – that is, powered dirigibles, high efficiency jets or electric propelled platforms that run off solar cells and fuel cells – offer improved spectrum efficiency. They support large phased array antenna platforms and achieve manifold frequency reuse with narrow beams generated closer to the earth's surface. HAPS allows for the possibility for hybrid satellite and HAPS networks that combine the relative strengths of satellites (broad area connective) with HAPS networks (high spectrum efficiency in a localized area).

SATELLITE SERVICES IN TRANSITION

Satellite evolution is supporting a wider range of services. For the past 20 years a process of specialization, allocation of frequencies for particular satellite services, and creation of organizations to provide services was sensible. However, digital processing and compression and convergence of information into a mega market made such processes less sensible. The situation was exacerbated by the emergence of the Internet, multimedia and the rapid global convergence of information, communication and entertainment companies. The system was a hindrance to future growth. That is, satellite technology has developed multipurpose platforms (in the form of very high power satellite buses), as restricting satellite systems to narrow frequency bands or narrow service categories would be likely to restrict growth and undercut the continuing decrease in service prices. Further, consolidated billing and administrative expenses would also allow cost reductions due to economies of scale.

Table 9.1 shows the current size of satellite markets by service category. Broadcast satellite services represent the dominant area of market expansion. What is important to note is that many broadband satellite systems planned for geosynchronous orbital operation are also designed to provide broadcast television, as well as broadband interactive multimedia services and rural and remote telephony. Conversely, those systems designed for broadband, interactive Internet-based services can also support direct-to-the-home television broadcast services. To sum, market demand has reshaped satellite system design, and frequency restrictions as set out in the frequency allocations of the International Telecommunication Union have been conveniently 'ignored' by satellite system operators. Another aspect of Table 9.1 is the extent to which there will be rapid growth of mobile satellite services. In spite of initial market failures by the low and medium earth

Table 9.1   Actual and projected satellite service revenue, 1992–2010 (USD billions)

Satellite service	1992	2002	2005	2010
Conventional fixed satellite service				
International and regional satellite systems	5.4	14.0	18.5	23.0
US/Canadian systems	2.3	4.2	6.0	7.0
Other national systems	1.4	4.5	6.5	8.0
All conventional fixed satellite systems	9.1	22.7	31.0	38.0
New broadband satellite system (Ku, Ka and Q/V Band)				
Broadband multimedia (GEO)	n.a.	1.0	4.0	9.0
Broadband multimedia (LEO)	n.a.	n.a.	3.0	6.5
All broadband multimedia systems	n.a.	1.0	9.5	14.5
Mobile satellite systems				
Mobile satellites (Aero / Maritime)*	0.8	2.0	2.5	3.5
Mobile satellites (Land / GEO)	0.01	1.0	1.5	3.0
Mobile satellites (Land / MEO / LEO)	n.a.	1.0	2.5	4.0
All mobile satellite systems	0.81	4.0	5.5	10.5
Broadcast (DBS / DTH) satellite systems	0.5	8.0	n.a.	45.0
Military satellite systems	n.a.	n.a.	n.a.	n.a.
Other (data relay / GPS / SNSS)	0.1	0.3	2.0	2.5
Total satellite service revenue	**10.5**	**36.0**	**77.0**	**110.5**

Notes:
International satellite revenues are rather estimated revenues of retail sales to customers.
International systems include INTELSAT, Cyberstar, PanAmSat, GE Americom, AsiaSat, Apstar, Eutelsat, Arabsat and the estimated international revenues of Hispasat, Optus and others.
* This is largely Inmarsat related and reflects not Inmarsat revenues but estimated total revenues derived from customers.

orbit (MEO) systems Iridium and ICO, the Globalstar system backed by Alcatel, Loral and others has shown rapid growth after deployment. The success of regional geosynchronous mobile systems such as Thuraya, ACeS and Agrani will become apparent in coming months. The demand for broadband, multimedia mobile satellite systems is the least well known.

Overall there seems to be a sustained market demand for broadcast radio and television, broadband, multimedia (especially those linked to the Internet), mobile satellite, rural and remote, messaging and disaster recovery services. As a combined, and increasingly integrated market, this group suggests strong future growth for the satellite industry. Revenue projections contained in Table 9.1 show satellite services exhibiting rapid growth. Satellite communications were expected to be worth USD27 billion for 2000 and projections are that satellite revenues will be approximately 3.3 per cent of global telecommunications revenues (that is, USD36 billion of USD1100 billion) at mid-2002 and approximately 5.4 per cent of global revenues by the end of 2005 (that is, USD77 billion of USD1450 billion). In projecting this continued market growth for satellite services it is important to note that not only will there be rapid growth of services in many sectors, but that the value-added aspect will continue even while terrestrial wireless values may show some decline. In short, satellites appear most likely to retain their market value in comparison to other telecommunications and information services that will continue to drop in price during the coming decade.

GLOBAL SHIFTS IN THE REGULATORY AND INSTITUTIONAL ENVIRONMENT

Some reasons why satellite system operators will be able to charge rates that retain their value derive from the regulatory and institutional environment. For several years satellite networks were merely treated as an extension of large telecommunications organizations that were often monopoly service providers. Satellite networks were mostly used when terrestrial networks were unavailable or overloaded. New satellite service providers offer different forms of competitive services and provide horizontal networks that directly interconnect the end user. This means that large-scale networks, even when there are more than 10000 nodes in the system, can be directly connected via very small aperture terminals or micro terminals known as ultra small aperture terminals (USAT). This means that satellite networks can provide television and movie entertainment, software, or broadband Internet services directly to end users.

Thus the customer can be directly billed for services on demand by a variety of satellite system operators. In this environment, satellite networks can eliminate intermediation by telecommunications and cable television operators. This arrangement allows the satellite operators and service providers to move much higher up the value chain. Satellite operators can combine the functions of entertainment and content provider, transmission and switching service provider, in addition to the functions of billing and marketing services. Furthermore, as the activities become less like the functions of a common carrier under strict control as to tariffs and more like those of an information network or entertainment service provider, these organizations will become less subject to regulatory control. Many of the organizations that are currently satellite service providers began operations as manufacturers of satellite hardware. Hughes Space and Communications is currently being sold to Boeing, leaving Hughes Communications as an entertainment and telecommunications services company. Loral, Alcatel Espace, Lockheed Martin, GE Americom and Matra-Marconi among others have entered the communications and information services markets, and in some cases (such as GE) have exited the satellite manufacturing industry.

ECONOMIC ISSUES

The complexity of satellite communications systems in terms of planning, engineering, implementation, financing, regulatory environment and continuing operation is much greater than that involved with fiber optic or coaxial and terrestrial wireless networks. Their theater of operation is broader, the regulatory environment is more extensive and demanding and the system is more complex and segmented. In short, many special challenges confront satellite system operators. Some of the elements of the complexity involved in satellite communications services include the need to obtain access to frequencies; the need to obtain and maintain international and national licenses to operate; long lead times to design, build and deploy satellite systems; the need to design and implement efficient systems whereby users can obtain access to satellite services directly through personal user terminals or the public switched telephone network (PSTN); the need to rapidly adapt to and implement satellite and ground system technology; special regulatory challenges to comply with licensing, tariffing and billing, and competitive constraints and challenges due to other service providers, that is, terrestrial and space based operators; maintaining compatibility with national and international technical standards and seamless interconnection with terrestrial cable, wire and wireless systems; and capital financing, launch insurance and risk management, and recruitment and training of staff.

No other form of telecommunications or information service provision has a greater challenge in terms of technology, regulatory and licensing constraints, longer term strategic and business planning, capital financing and difficulty of business operations. Factors that impact the cost of providing satellite service and in turn service pricing are manifold. Some factors vary from service to service or are peculiar to localized markets, such as localized zoning constraints, official corruption, and access to reliable or low cost electric power. The following issues are global concerns that apply to virtually all satellite service providers and will serve as major cost and price drivers in the decades ahead. The critical asset that will dictate the future rapid growth of satellite communications is radio frequency spectrum. New higher radio frequency bands have been allocated. Satellites have been more closely spaced together in geosynchronous orbit and new LEO and MEO systems have been deployed. Further, new digital compression techniques and ways to allow frequency bands to be reused have helped to send more information through available spectrum. The rising demand for satellite-based spectrum, especially for broadband and mobile services, has pushed this issue into prominence.

The current value of a satellite slot in geosynchronous orbit has risen to millions of USD per annum. Technical solutions will provide greater access to additional spectrum, but use of the higher frequencies will, at least for the next ten years, command higher cost. The reallocation of spectrum for satellite use will not be easily achieved and will involve terrestrial broadband wireless alternatives. The advent of HAPS will only complicate this issue, and to better understand it metrics that measure the cost of satellite spectrum are needed. Measures of the net cost per Hz per annum in different bands and in different orbital configurations, as well as the cost of satellite spectrum per KHz-Kilobit/sec over time are required. An apparent trend is that the net cost of raw spectrum is rising, but empirical studies are needed to document these trends. A key is to measure the extent to which new technology and regulatory actions can allow the net cost of satellite spectrum as measured in KHz-Kilobit/sec to decline.

Satellite services can be delivered either as mass media and consumer oriented services that go directly to end users, or satellites can serve as an extension of the PSTN. In the latter case satellites act as an adjunct to a larger network and are placed much lower down the value chain. That is, satellite systems that operate via PTSN gateways are usually only a secondary transmission medium, do not have their own billing systems and receive a very small portion of the revenues earned for the service provided. Since the mid-1990s satellites have made the transition from being a gateway-based service that indirectly serves consumers to an end user oriented service. This transition began with satellites being able to offer services directly to end users for entertainment video and computer games. Direct broadcast or direct-to-the-home services opened new markets to satellite service providers. By mid-2000, there were an estimated 50 million direct-to-home television subscribers globally (15 million in Japan, 12.5 million in the US and 27 million in Europe and elsewhere). Direct audio broadcasting was beginning full operation via such organizations as Worldspace, XM Broadcasting and the Sirius satellite system, with the number of subscribers projected to reach 10 million by 2005.

Mobile satellite services via a combination of geosynchronous, MEO and LEO systems are the second wave of this move toward the satellite industry becoming a mass consumer market service. Closely linked to the mobile satellite services are the over 1.5 million consumer users of GPS navigation systems. This number is also expected to reach 10 million by the mid-2000s. Currently several integrated satellite tracking and messaging satellite systems – Omnitracs, Euteltracs and Orbcomm – are developing this market. By the end of 2001 the Hertz car rental company was due to have 1 000 000 'never lost' units installed in their global fleet that combine services from Orbcomm and GPS. The next wave of the satellite revolution to become a mass consumer type service will occur in the 2002 to 2006 period. The objective is to provide new broadband satellite services designed to beam Internet-based, multimedia services to the desk top via USAT or to mobile units. These fixed interactive microterminals will be 50 cm to 65 cm in size and able to provide a wide range of business services on demand at between 4.8 kilobit/second to several megabits/second. These broadband systems will operate in the Ka and Ku band frequencies and eventually will migrate to the higher Q/V bands in the 2010s. Although satellite systems will still interconnect via major gateways, an increasing portion of global satellite revenues will be via small terminals that provide service to end users. This basic change in the scope, nature and indeed the size of the satellite market will imply a host of other changes in the way the satellite industry operates, collects its bills, provides security to its users, and accesses national and international markets. The most important of the satellite industry concerns during the next decade are summarized in Table 9.2.

Should current trends continue the distinction between satellite service providers may very well continue to diminish as multipurpose platforms emerge to offer a complete suite of multimedia services that include television, electronic games, telephone, facsimile, e-mail and broadband Internet access. While currently there are different satellite and ground antenna systems used to provide service, similarities are beginning to emerge. Further, institutional reforms and the privatization of INTELSAT, INMARSAT and EUTELSAT will also bring further unity to pricing and satellite competition. Convergence of satellite services into integrated systems based on multipurpose systems is still some time away. Some specialization such as direct broadcast radio and messaging services via satellite could remain for the longer term. This continuation of these types of satellites would be heavily based on the many millions of specialized and low cost user terminals that are in general use in automobiles, trucking and shipping fleets, pipeline networks and so on. But for DBS/DTH, for mobile satellite services, and for broadband, multimedia and Internet satellite

Table 9.2   Major challenges to the global satellite industry

Nature of the challenge	Types of solutions
Security	Improved encryption and subscriber identity.
Access to national markets	WTO protection through GATS, competitive entry support within the EU, bilateral trade agreements, and improved collection and settlement policies.
Frequency allocation	Improved flexibility in general allocation of frequency for multiple use, incentive to improve frequency reuse, hybrid satellite and HAPS stratospheric platforms.
Privacy	Improved encoding and encryption systems, protection in mobile services billing systems, and national legislation and international guidelines.
Intellectual property and programming	Collection of copyright fees at distribution point, WIPO regulations, e-streaming vs electronic cache updating, bridging ownership in satellite systems and program production.
Direct billing and bypass of national gateways	National legislation and international guidelines, WTO provisions, enforcement by EU and regional organizations.

services, convergence in pricing and customer oriented service provision is appearing. For instance most satellite services can be obtained on a short-term basis such as per minute charge, or on a medium-term basis of weeks or months, or on a wholesale basis for a longer term of several years. Ownership of satellite capacity or the equivalent of indefeasible right of use or capacity can be obtained through a resale market from brokers for whatever term desired. In this new satellite industry the general public or business can obtain ownership and use of small customer premises terminals and pay for satellite usage on a monthly lease basis, much like one would obtain cable television or telephone service. The most interesting economic and pricing issues do not divide along service lines, but rather along the lines of service options and special areas of user concerns.

The more important pricing and service issues for satellite telecommunications to emerge over the next decade include the following.

(a)   *Tariffs versus open pricing:* Initially satellite communications evolved under strict regulatory control with fixed and published tariffs provided exclusively through telecommunications carriers on a monopoly basis. As a result of swift global regulatory transition this is less the case. Although some satellite offerings are based on tariffs that are set under rate base or price caps with incentives regulatory systems, the new environment increasingly reflects open and competitive market conditions. This less regulated environment provides more opportunity to offer value-added services and greater price flexibility. This includes versioning of services with premiums charged for greater reliability, quality of service, enhanced audio, access to improved software or programming, priority, and interactive messaging via pagers.

(b) *Satellites, Internet and IP networking:* The Internet and corporate intranets have fundamentally changed telecommunications and networking. Satellites must be able to handle traditional telecommunications and the PSTN by supporting number seven signaling and ATM, but must also provide quality service to support Internet and private IP networks. Since private IP networks operate at low cost (as low as USD0.001 per Mb of throughput) this represents the greatest challenge to modern satellite networks. Satellite systems are especially challenged to meet quality of service, and to overcome transmission latency problems and the conversion from WDMA to TDMA or CDMA multiplexing. In short, satellite system operators are challenged in terms of price performance and with being flexible enough to support X.25, frame relay, SMDS, ATM, TCP/IP and IP over ATM with a minimum of technical reconfiguration.

(c) *Satellite versus fiber, cable modem and digital subscriber loop:* The complexion of future direct consumer markets will depend on satellite systems' ability to compete with hybrid fiber coaxial, cable modems, broadband terrestrial wireless and digital subscriber loop. The key to satellite competitiveness involves the ability to develop and install compact and low cost micro terminals that fall in the range USD250 to USD1000. Of course the satellite network in space will need to increase in power and perform as cost per Mb decreases; however, should the ground segment not approach these threshold prices then advances by the space segment will be irrelevant. In short, satellite markets will be restricted to broadcast, mobile, and rural and remote services.

(d) *Security, enhanced information processing and value-added services:* The future for fiber optic systems, supported by advanced wave division multiplexing, defines an advanced communication environment where broadband services are available with a high quality of service, reliability and low cost. However, providers will be challenged by the users increasingly demanding improved privacy and security systems, improved and more user-friendly billing systems, and a host of other value-added capabilities. Networking and computer software companies that have had much greater experience and the opportunity to focus on value-added services clearly have an advantage over satellite system providers. While it is widely assumed that fiber networks offer greater security over satellite and wireless systems, improved encryption can serve to overcome some of these liabilities.

(e) *Market access:* Despite major improvements in market access for satellite services, DBS/DTH, mobile, and conventional fixed satellite services there remain barriers to national markets. Additional reforms through the WTO and GATS agreements, open trade enforcement, pressure through the ITU on collection and settlement rates, and improved opportunity for user ownership of consumer premises antennas can make a difference. That satellites represent a bypass technology threatening national telecommunications entities clearly underscores the industry's vulnerability compared to terrestrial telecommunications systems. In many national markets, even when satellites are price competitive they face problems due to their special regulatory status and bypass architecture.

(f) *Intellectual property, WIPO and copyright issues:* As the cost of telecommunications transmission via media, whether fiber, coaxial, terrestrial wireless or satellite, continues to drop toward zero, and as data transmission becomes a commodity the importance of software and content will increase. Eventually consumers will pay higher prices to block unwanted information for transmission time to receive data. Currently, it can cost 50 to 100 times more to print incoming information than to receive it. Value-added telecommunications services from terrestrial or satellite service providers to protect proprietary data and copyright for movies or television programs will dominate simple information transport.

(g) *Merger of communications, geomatics and navigational services:* The integration of broadcast entertainment services with interactive telecommunication services, earth imaging and navigation services will become an increasingly important market during the next decade. The Internet, e-streaming entertainment, open systems and group ware, web mapping, intelligent highways, geological applications, military reconnaissance and intelligence services will reinforce the need to allow users globally to combine electronic applications into integrated systems. These services will eventually be available through laptops, palmtops or wearable units.

CONCLUSIONS

Satellite communications growth will be driven by many factors such as price and technical competition from fiber and other transmission technology, regulatory and trade reform, and advances in frequency allocation. Satellite communications will remain a complement to the predominant fiber and terrestrial wireless systems, including new stratospheric platforms, and provide broadcast, mobile, networking and rural and remote interconnection to supplement ground-based systems and offer special value-added services. The continued decrease in price for telecommunications services and the increase in derived bandwidth, via intensive frequency reuse techniques, to support multimedia and multicasting services will be the same in space and on the ground. This means that the demand for satellite services and their pricing will be in an intensively competitive market. Developing countries will have special needs and major growth opportunities for satellite system operators, but satellite systems will fill niches in the broadband fiber networks across the developed economies. The most important factors affecting the growth of satellite communications include the development of USATs or micro terminals that can be sold at USD1000; the ability to access additional RF spectrum and find cost effective ways to derive frequency through intensive reuse techniques; complementing fiber networks rather than being a direct competitor; growth of terrestrial mobile and wireless systems that will shrink the market opportunities for land mobile satellite systems; geosynchronous satellite systems that will for both cost and technical reasons dominate LEO satellite systems, while the emergence of stratospheric platforms will reduce LEO system opportunities further; and value-added services and Internet access to customers. Equally so, the move from tightly regulated and tariffed services to market-based and competitive pricing and the effective use of versioning in service offerings; provision of flexible and effective support to both PSTN and Internet/intranet enterprise networks; and integration of space imaging, navigation, telecommunications and broadcast satellite systems in terms of single, unified user units will be a key service demand delivered on a web-based platform; and the development of hybrid satellite and stratospheric platforms may represent one of the most important strategies to obtain access to additional frequency.

BIBLIOGRAPHY

Burton, E. and Hyde, G. (1998), *Laser Satellite Communications, Programs, Technology and Applications*, Washington, DC: IEE-USA Aerospace Policy Committee, April.

Evans, J. (1998), 'New satellites for personal communications', *Scientific American*, 278(4), 70–79.

Forrest, J.R. (2000), 'Communications networks for the new millennium', *Intermedia*, 28(1), 24–8.

Futron, (2000), 'Satellite Industry Association, and the Space Policy Institute', *Satellite Industry Guide*, Washington, DC.

Hartshorn, D. (2000), 'Satellite communications: You can't keep a good technology down', *Communications Technology Decisions*, 1, 111–14.

Lamberton, D. (ed.) (1997), *The Economics of Communications and Information*, Cheltenham: Edward Elgar.

Pelton, J.N. (1998), 'The future of 21st century telecommunications', *Scientific American*, 278(4), 80–85.

Pelton, J.N. and MacRae, A. (1999), *WTEC Panel Report on Global Satellite Communications Technology and Systems*, Baltimore, MD: ITRI.

Shapiro, C. and Varian, H.R. (1999), *Information Rules*, Cambridge, MA: Harvard Business School Press.

Subcommittee on Computing, Information, and Communications R&D, National Science and Technology Council, OSTP, The White House (1999), *Information Technology Frontiers for a New Millennium* (Second Printing), Washington, DC.

White, N. and White, L.J. (1998), 'One-way networks, two-way networks, comparability and antitrust', in Gabel, D. and Weiman, D. (eds) *Opening Networks to Competition*, Boston: Kluwer Academic Press.

10. Regulated costs and prices in telecommunications

Jerry A. Hausman

INTRODUCTION

Economic advice to regulators regarding the correct principles to set regulated prices has often been flawed in that it does not recognize the underlying technology of the industry. Economists recognized early on that in the situation of privately owned utilities in the United States (US) the first-best prescription of price set equal to marginal cost could not be used because of the substantial fixed (and common) costs that most regulated utilities needed to pay (see Kahn, 1970). This realization typically accompanied the claim that the economies of scale of the regulated firm were so significant that competition could not take place because the regulated firm's cost function was significantly below that of new entrants. Nevertheless, the most common advice from economists was that prices should be set similar to the outcome of a competitive process. What the competitive process would be was never specified in any detail, which was to be expected since economic theory had no well-accepted model of competition with a technology exhibiting strong economies of scale, especially in the multi-product situation.

In the US, regulators following legal principles adopted the position that the regulated firm should cover its costs. However, regulators also adopted prices for certain services to attempt to meet social goals for these given services. For other services, regulators used arbitrary means to set prices while balancing competing claims from increasingly well organized groups of consumers, all of whom claimed they should receive low prices with other groups paying for the fixed and common costs. This regulatory approach arguably did not do undue damage when no actual competition existed. So long as the regulated firm was (nearly) productively efficient, the losses were essentially second-order social welfare losses.[1] The regulated firm covered its total costs, at least approximately, although prices for individual services were often badly distorted from an economically efficient solution. However, when actual competition appeared and was allowed to exist by the regulators, the economists' advice of setting prices as if they were the outcome of a competitive process soon led to a regulatory morass. Regulators could no longer depend only on cost factors in setting regulated prices. The outcome of a competitive process would also need to take into account demand and competitive interaction (oligopoly) factors, with the first set of factors difficult to measure and the competitive interaction factors unlikely to be agreed on. While regulators had some imperfect information about costs, they typically had little or no information about demand and no well-developed idea regarding the effects of competitive factors. In the following two sections the question of whether using costs to set regulated prices, while disregarding demand factors and competitive factors, is a reasonable economic policy is discussed.

A particularly difficult problem arises when a regulated firm wants to decrease its prices for services subject to entrant competition. Economists recognize that price set above incremental (marginal) cost should be permitted. New entrants want the previously regulator-set prices to be maintained. New entrants typically enter because regulated prices are well above efficient levels, and they do not want these prices reduced. Furthermore, from a social welfare viewpoint the argument became first order since inefficient new firms could be productively inefficient, causing a first-order loss of social welfare. Regulators found it difficult to permit the regulated firm to reduce its prices, since under cost of service regulation other prices would need to increase. Even when cost of service regulation was replaced by incentive (price-cap) regulation in the 1980s and 1990s, regulators found it extremely difficult to allow price reductions since they believed in 'regulated competition' (an oxymoron) where regulators could better manage competition than the market. Nevertheless, the regulated companies were not harmed too badly since competition did not proceed at such a rapid pace as to cause extreme economic damage.

Cost-based regulation of telecommunications – for example rate-of-return (ROR) regulation in the US – had substantial negative effects on innovation while it was claimed that it led to excessive capital investment. Most economists conclude that cost-based regulation led to significant consumer harm. In the mid-1980s when the UK government privatized British Telecom, it decided not to use the historic approach of cost of service regulation to set regulated prices as the US and Canada had. The UK government instead chose price caps – a regulatory method proposed by Littlechild and discussed in Beesley and Littlechild (1989). Price caps are regulated prices based on inflation and a productivity factor, instead of regulated profits as in the US cost of service-based ROR regulation. Price caps have a number of advantages over ROR regulation in terms of incentives for cost minimization (productive efficiency), innovation, and the ability of the regulated firm to rebalance its prices. In particular, the regulated firm can reduce its prices to compete. In 1989–90 the US Federal Communications Commission (FCC) adopted price caps. During the 1980s and 1990s price-cap regulation was implemented instead of cost-based regulation in most countries when telephone companies and other utilities were privatized. In the majority of the states of the US, ROR regulation has been replaced by price-cap regulation. The battle to banish cost-based regulation appeared to be largely over.[2]

During the late 1990s and early 2000s cost-based regulation has reappeared because of the necessity to set a price for unbundled network elements sold by incumbent firms to their competitors. Several governments including the US, Australia and Canada adopted mandatory network unbundling for the incumbent local exchange carrier (ILEC). The most commonly used approach to set regulated network element prices was based on total service long-run incremental cost or TSLRIC. Unfortunately, the adoption of TSLRIC as a cost basis to set the prices for unbundled elements has negative economic incentive effects for innovation and for new investment in telecommunications networks. TSLRIC provides an incorrect basis on which to set regulated prices as it fails to recognize that a significant proportion of telecommunications networks are sunk costs. Instead, TSLRIC makes the assumption that costs are fixed but not sunk so that the capital assets could be redeployed in other uses if technology advances or other economics events decreased the return on the assets. Failure to recognize the sunk-cost character of much network investment leads to the granting of a free option to the regulated incumbent's competitors. This causes incumbent firm shareholders to fund the free option, as competition will lead to under-investment by both the incumbent and competitors. The incumbent under-invests because it will not achieve (on average) a sufficient return to justify marginal investments due to the grant of the free option to its competitors. New competitors, who receive the 'free option', will under-invest in facilities because of the subsidy they receive with the grant of the free option. Uncertainty in a dynamic industry, with rapidly changing technology and economics, can have an especially large effect on investment incentives because the value of the option is high. Losers are consumers and businesses that do not have access to the most up to date services, had the regulation not created disincentives to invest.

How did network unbundling and a return to cost-based regulation become government policy? In 1996 the US Congress passed the Telecommunications Act. As a trade-off for permitting local telephone companies to provide long distance services, they agreed to unbundle their networks.[3] The FCC adopted cost-of-service regulation to set the unbundled network element prices. Thus the well-known problems of cost-of-service regulation with its inability to correctly treat economies of scale and scope, and its use of arbitrary allocations of fixed and common costs to prices all reappeared. Even worse, the FCC adopted the approach of total element long-run incremental cost (TELRIC) that assumes that all investments in telecommunications networks are fixed, but not sunk. This assumption is, of course, directly contradicted by the actual technology of telecommunications networks. Perhaps an even more troubling development is that a number of countries such as the UK and Australia have adopted a similar incorrect regulatory cost-based approach called total service long-run incremental cost (TSLRIC). It appears likely that the European Union will adopt a similarly incorrect approach. What is particularly troublesome is that the inventor of TSLRIC has now stated that the failure to account for sunk costs is a mistake.

Below I discuss why the cost-based approach to regulation, which ignores demand factors and competitive factors, is wrong except under a very special set of assumptions. The assumptions, used in the 'non-substitution theorem' are closely connected to Marx's labor theory of value, and never hold true, even approximately, in real-world telecommunications networks. Thus the regulatory attempt to set prices independent of demand does not make economic sense. However, even within this approach, why the failure to take account of sunk costs leads to a large downward bias in setting regulated prices is discussed.[4] The assumption that network investments are fixed, but not sunk, leads to a large error. Also, by giving a 'free option' to new entrants the policy creates an economic disincentive for facilities-based investment by the new entrants. Instead, they find it better to accept the below-cost use of the incumbent provider's network. Thus the regulators' attempt to set price that would occur in a competitive market is very far removed from the real-world technology and competition that would exist in a competitive telecommunications market. FCC-type regulation is leading to reductions in economic efficiency and decreased consumer welfare. Instead, regulators should permit actual competition to occur rather than trying to choose the form of regulated competition they think should take place.

In the final section of this review chapter I consider the question of which elements of the incumbent's network should be subject to mandatory unbundling.[5] The goal of the 1996 Telecommunications Act is increased consumer welfare and competition. Thus a consumer welfare approach to mandatory unbundling is discussed. The approach is in contrast to the US regulators' approach of a competitive welfare standard. A competitor welfare approach leads to reduced investment and innovation compared to a consumer welfare approach. The likely outcome of government policy in the US, in contrast to the approach taken in Canada and Australia, will be to harm US consumers.

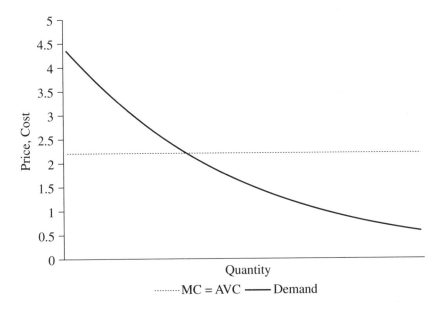

Figure 10.1   Cost and price with constant returns to scale

A SIMPLE MODEL OF COST-BASED REGULATION

The model of cost-based regulation is to use costs of production to set prices that would be the result of a 'competitive' situation. These costs of production are used to set prices independent of demand factors. A very simple one-good one-period Marshallian partial equilibrium model leads to the result, where competitive price is independent of demand.

Conditions for Prices Independent of Demand

Assume that a regulated telecommunications service is produced by one or more inputs. No multi-period capital goods are present. The production technology exhibits constant returns to scale. In Figure 10.1 it follows that competitive price equals marginal cost, which in turns equals average cost, because of the constant returns to scale assumption. The position and shape of the demand curve does not matter in setting the competitive price. Under these conditions, cost determines price, independent of demand. This interesting result depends very much on the assumptions of the economic model: partial equilibrium so that demand for the product does not affect input factor prices, constant returns to scale so there are no economies of scale, a single product so there is no joint production and no economies of scope, and a single period so there are no durable capital goods. When any of the assumptions fails to hold, competitive price cannot be based on cost, independent of demand. Thus the price independent of demand result turns out to be a very special result not applicable to real-world telecommunications.

The Role of Fixed Costs and Economies of Scale

Now suppose that marginal cost remains constant but that fixed costs of production are introduced. However, a single service continues being produced. The cost function is written as:

$$C(q; w) = F + wq \qquad (10.1)$$

where F is the fixed cost, q is output quantity, and w is the constant marginal cost per unit of output. A regulator might conclude that in a competitive, free entry situation price would equal average cost, so that $p = (C/q) = (F/q) + w$. Since quantity demanded is a function of price, price is no longer independent of demand. However, setting price equal to average cost, AVC, seems to be the correct outcome if the regulated utility is to recover its costs.

The Role of Common Costs and Economies of Scope

A common cost arises when more than one service derives from a production process, but some of the cost is incremental to neither product. The term 'fixed and common costs' arises often in discussion of regulated costs and prices because of the common occurrence of this type of cost. In terms of the cost function, again assume constant marginal costs for both outputs:

$$C(q_1, q_2; w_1, w_2) = G + w_1 q_1 + w_2 q_2 \qquad (10.2)$$

Note in (10.2) the fixed cost G cannot be uniquely assigned to either output. Indeed, no measure of average cost for either output exists. Here regulators typically choose to use an allocation of the fixed cost G to each service. However, allocations such as fully allocated cost, equal allocation of cost and so on are inherently arbitrary.[6] Nevertheless, the results of such allocations have very important consequences for regulated prices. These regulated prices in turn have important effects on competition, economic efficiency and consumer welfare. In competitive markets, firms set price based on cost, demand and competitive conditions. Regulators attempt to base prices only on the first of these factors. Thus regulators do not meet their goal of setting regulated prices in a similar manner to competitive markets. Furthermore, they cause billions of dollars annually in losses in economic efficiency and consumer welfare.[7] Instead of using inherently arbitrary allocation procedures, regulators should either take account of demand and competitive conditions in setting regulated prices or adopt procedures such as global price caps which will lead the regulated utility to take account of demand and competitive conditions.[8]

The Role of Sunk Costs

The model is generalized further by considering sunk costs in addition to fixed costs. Sunk costs are costs that cannot be recovered when the economic activity ceases. Sunk costs are prevalent in telecommunications networks – for example, consider an investment in a (copper) loop to a residential customer. The customer has a unique loop that connects the residence to the central office switch. When this customer decides to use a competitive service, such as a local access service offered by a competitive cable company or a wireless company, the copper loop cannot be redeployed. The investment in the loop is sunk. When a regulated telephone company faces no uncertainty over the future use of the loop and the cost and prices for the associated services provided with the loop, the distinction between a fixed cost which arises from an asset which can be economically redeployed and a sunk cost is not that important. Indeed, in the 'old days' of cost-based regulation for a monopoly provider, if an investment was deemed to be 'used and useful' by the regulator, the asset entered the regulatory cost base. Once the asset entered the regulatory cost base, the regulator, in principle, allowed the utility to recover the cost of the investment.[9]

However, in a situation of competition, where the utilities' competitors are allowed to use the incumbent's network at regulated prices, the distinction between fixed and sunk costs can be quite important. The competitor typically pays for the facility it uses on a monthly basis. As regulators assume investment costs are fixed but not sunk, competitors receive a free option to use the incumbent's network at a price that fails to take account of the sunk cost nature of much of the investment. The regulators thus subsidize competitors at the expense of the incumbent and create an economic disincentive for competitors to invest in their own facilities.[10] Furthermore, the regulators reduce the incentive for new services to be offered by the incumbent. New services often fail. Yet if successful new services must be resold to competitors at cost, the incentive to undertake the required risky investment is diminished.[11]

COST-BASED REGULATION: ECONOMIC ANALYSIS WITH COST BUT NOT DEMAND

In a simple one-period and one-good production model with constant returns to scale, a partial equilibrium Marshallian analysis demonstrates that the competitive price does not depend on demand. Marginal and average costs are independent of the quantity produced, so the position of the demand curve does not affect the price. However, the required description of technology does not depict accurately the telecommunications industry. For example, telephone and wireless networks have a very large proportion of fixed and sunk costs. Whether the 'price independence of demand' type result holds in a broader context is considered next to see if this result is (approximately) applicable to telecommunications. To do so, non-substitution theorems – which demonstrate that under certain conditions an economy will have a unique price structure determined by the costs of production and independent of the structure of final demand – are considered. These results are referred to as Samuelson–Mirrlees non-substitution theorems (see Samuelson, 1961 and Mirrlees, 1969).[12] Initially consider only the simplest situation where labor is the only non-produced factor in the economy. Here a set of necessary conditions that would lead to a Samuelson–Mirrlees non-substitution theorem result. These necessary conditions are the following.

(a) *Only one non-produced good exists.* The good is usually assumed to be labor so that land or minerals do not exist.
(b) *The technology has constant returns to scale.* A constant per unit requirement of inputs occurs regardless of the amount of output. This condition rules out economies of scale.
(c) *No joint production.* A single production process cannot lead to two or more different outputs. This condition rules out economies of scope.
(d) *The economy is productive.* The economy can produce a positive net vector of outputs where net output is gross output minus inputs.

With these (plus some additional technical) conditions, product prices are independent of final demand. Product prices equal the cost of production, denominated in terms of the *numeraire* that can be units of the non-produced good. Thus, in a Samuelson–Mirrlees non-substitution model, prices of the many products in the economy are independent of demand, as in the simple partial equilibrium single-product Marshallian model.

Enter the Marxian Theory of Value

Since labor is the only primary input in an economy described by the non-substitution theorems and prices are independent of demand, what sets this price? Prices are set by the cost of production, as in the Marshallian example, and the cost of production is the sum of direct plus indirect labor costs in a one-period economy.[13] Actually, solving the dual problem to the linear programming problem (which minimizes the cost for a given final output vector that yields the non-substitution theorem result) leads to the conclusion that the labor costs will be minimized in the problem. These minimized costs establish the prices in the non-substitution theorem economy and are independent of final demand. This result is similar to the Marxian labor theory of value (see Morishima, 1973). When the situation is generalized to more than one period and durable capital goods are present, the cost of production remains direct plus indirect labor costs. However, the labor costs embedded in the durable capital goods increase at the economy-wide rate of interest, connected to the steady-state growth rate of the economy, each period.

It is worth noting that the 'Marxian theory of value' terminology is not a 'Marxian theory of price'. The Marxian theory of value arises from the labor cost of production theory as discussed above in a particular multi-sector economic model. A huge literature exists that attempts to go from this labor theory of value to the competitive price in the context of Marxian analysis – the so-called transformation problem between values and competitive prices (see Samuelson, 1971 and Morishima, 1973). Marx understood that market determined competitive prices could differ greatly from those of the labor theory of value.[14] Furthermore, Marx and his followers were unsuccessful in solving the transformation problem except under very restrictive and uninteresting assumptions. Thus both Marx and his followers were unable to go from a cost basis in terms of labor costs to observed competitive prices (independent of demand). Cost-based regulation is involved in a similar exercise to this 'crude' Marxian economics of determining prices that would result in a competitive economy solely from some measure of cost, which is an impossible task under realistic economic conditions. But the regulators' attempt to set competitive prices while disregarding demand has some interesting connections with Marxian economic analysis. Regulators and some Marxian economists have attempted a remarkably similar yet mistaken approach to determine competitive prices from a basis determined solely by the costs of production.

Necessary Assumptions and Economic Reality: The 'Regulatory Fallacy'

Could the regulatory goal of setting competitive prices independent of demand hold approximately true in a realistic economic situation? Since the assumptions for the Samuelson–Mirrlees non-substitution theorems are necessary assumptions, no weaker ones will do. Thus, to correctly set prices independent of demand, the four necessary assumptions must hold true. The first assumption of only a single non-produced factor cannot be correct in a modern economy. If labor and land (minerals) are both non-produced factors their relative prices will affect input costs and final product prices. But their relative prices will depend on the pattern of demand for products that use both labor and land (silicon, copper and silver). Since products will use in direct and indirect form different proportions of the non-produced factors, the relative prices cannot be independent of demand.[15] Then neither the cost of production nor final product prices can be independent of demand. How important this departure from the necessary assumption is cannot be resolved easily. It may not be that important since, should it be considered that telecommunications is a separable sector of the economy – similar to partial equilibrium analysis – it might be claimed that the sector is small enough compared to a given regional economy for service and the world economy for capital goods that it does not have a significant effect on the relative prices of primary factors. The price of the Hicksian composite economy for the non-telecommunications sector might be used as a *numeraire* without too much departure from reality. The last assumption, that the economy is productive, is disposed of with the remark that as an approximation its likely departure is unimportant.

The most important necessary assumptions for the current application are no economies of scale and scope. The presence of substantial economies of scale has traditionally been given as one of the primary reasons for regulation (see Kahn, 1988, II, pp. 119ff.). The old question of a natural monopoly is based on large economies of scale. Whether or not the claim of a natural monopoly is correct, modern telecommunications network regulation in the US, the UK, Australia and Canada is based on the importance of economies of scale.[16] The idea is that a new entrant cannot duplicate the telecommunications network so the incumbent provider is required to sell the use of its network to the entrant at regulated cost. The common terminology of fixed and common costs in telecommunications denotes the importance of economies of scale that arise from fixed costs in modern telecommunications networks. The regulated price typically ignores demand factors that are inconsistent with the notion of economies of scale. The higher the demand the lower the per unit cost, especially when fixed costs are taken into account.[17]

The no economies of scope assumption of the Samuelson–Mirrlees non-substitution theorems is violated by all modern telecommunications networks. An example of joint production arises with modern telecommunications switches, which are combinations of computers and switch blocks.[18] Switches route calls but they also provide other services such as voice mail. The same computer is used to provide both services in a less costly manner than if switching and voice mail were provided separately. Again economies of scope are one of the stated reasons for required resale of network functions by incumbent telephone companies to their competitors. Another indication of the importance of economies of scope is the concern with common costs in debates over regulated prices. Common costs are typically defined to be costs that arise from two (or more) services, but the costs are not incrementally caused by either service alone. The FCC, the Canadian CRTC, and some state regulatory bodies have arbitrarily set a mark-up to the direct cost of 20–25 per cent to take account of common costs.

Yet economists know that most modern competitive companies have joint production and common costs for the production of their outputs. These companies base their prices on competitive conditions for their products. Competitive conditions take account of demand conditions that arise from overall market demand for the product as well as firm demand conditions that arise as a result of competition. While regulators often say they want to replicate the outcome of a competitive process, they miss the obvious point that a competitive process involves cost factors as well as demand factors. Regulators, on the contrary, ignore the effect of demand factors on competitive outcomes, instead they use arbitrary mark-ups over some measure of incremental (or variable) cost to account for economies of scale and scope. These arbitrary mark-ups decrease economic efficiency and consumer welfare substantially.

An additional necessary assumption for a non-substitution theorem to hold is that the economy is on a steady-state growth path. This assumption allows for durable capital goods to enter the model. This assumption for an economy may be a reasonable approximation in certain circumstances, but for the telecommunications sector it departs from an approximation to economic reality.[19] Economists agree that the telecommunications sector is among the most dynamic in the economy. And since the durable capital goods used in the telecommunications sector are closely connected to semiconductors and optical transmission, innovations in these sectors will directly affect investment in capital goods in telecommunications. Thus the steady-state growth assumption is not a good assumption for telecommunications.

Hence my overall evaluation is that modern telecommunications differ in many significant and quantitatively important ways from the necessary conditions for price to be independent of demand. Economies of scale and scope are universally recognized to be important economic characteristics of modern telecommunications networks. The regulatory attempt to set prices as if they were the outcome of a competitive process but to ignore the importance of demand factors leads to the 'regulatory fallacy'. No serious student of economics would claim that the necessary conditions for the non-substitution theorem hold in a telecommunication network environment. Yet the regulatory assumption that price would be based on cost alone in a competitive market is wrong. Economic theory has developed precise conditions when price is independent of demand, and they do not hold, even as an approximation, in telecommunications. Thus regulators are acting on an erroneous belief that, with competition, price equals cost, independent of demand. This erroneous belief leads directly to the resulting regulatory fallacy. The regulatory fallacy leads to the consequent use of arbitrary allocations and mark-ups to regulated prices to take account of fixed and common costs. These costs are exactly the costs that arise from economies of scale and scope. The regulatory approach leads to significant consumer harm. If regulators instead took account of demand factors in setting regulated prices, economic efficiency and consumer welfare could be increased significantly.[20]

FIXED AND SUNK COSTS IN COST-BASED REGULATION

Current FCC Approach to Regulation of Unbundled Elements

The US Congress passed the Telecommunications Act of 1996, which called for less regulation, more competition and the most modern up to date telecommunications infrastructure: '[T]o provide for a pro-competitive, deregulatory national policy framework designed to accelerate rapidly private sector deployment of advanced telecommunications and information technologies and services to all Americans by opening all telecommunications markets to competition'.[21] The FCC instituted numerous regulatory rulemakings to implement the 1996 Telecommunication Act. The most important regulations so far have been the Local Competition and Interconnection Order of August 1996.[22] If implemented in its current form the Order is likely to have serious negative effects on innovation and new investment in the local telephone network.[23]

Most economists agree that regulation should be used only when significant market power can lead to unregulated prices well above competitive levels.[24] The goal of regulators is then to set prices at 'competitive levels'. However, economists are much less explicit about how these competitive price levels can be estimated. Most economists would agree that perfect competition cannot yield the appropriate standard since prices set at marginal cost will not allow a privately owned utility to earn a sufficient return on capital to survive. The large fixed costs of telecommunications networks thus do not allow the 'price equals marginal cost' standard of perfect competition to be used.[25] Baumol and Sidak (1994) proposed the perfect contestability standard as an alternative, where regulators should require firms to set prices as if 'the competitive pressures generated by fully unimpeded and costless entry and exit, contrary to fact, were to prevail'.[26] However, costless entry and exit presumes that no sunk costs exist – that is, costs that cannot be recovered on exit by a firm.[27] This assumption of no sunk costs is extremely far from economic and technological reality in telecommunications where the essence of most investments is an extremely high proportion of sunk costs. Consider the investment by an ILEC in a new local fiber optic network that can provide broadband services and high speed Internet access to residential customers. Most of the investment is sunk since if the broadband network does not succeed, the investment cannot be recovered. Thus, when either technological or economic uncertainty exists, 'perfect contestability as a generalization of perfect competition' cannot provide the correct competitive standard.

In a perfectly contestable market, if the return to an investment is below the competitive return, the investment is immediately removed from the market and used elsewhere. This costless exit strategy is always available in a perfectly contestable market.[28] However, the actual economics of telecommunications investment could not be further from a perfectly contestable market. When fiber optic networks are constructed, they are in large part sunk investments.[29] If their economic return falls below competitive levels, the firm cannot shift them to other uses because of their sunk and irreversible nature.[30] Thus the use of a perfectly contestable market standard fails to recognize the important feature of sunk and irreversible investments – they eliminate costless exit. Because of its failure to take into account the sunk and irreversible nature of much telecommunications investment, the contestable market model has nothing of interest to say about competition in telecommunications.[31] An industry cannot be expected to behave in a manner that is fundamentally inconsistent with its underlying technological and economic characteristics.

One way to consider the problem is the situation of a new investment by an ILEC. Suppose a competitor wants to buy the unbundled elements associated with the investment. The ILEC could offer the new competitor a contract for the economic life of the investment – say ten years for investment in the local loop. The price of the unbundled element would be the total investment cost plus the annual operating costs for the unbundled element. Should demand not materialize or prices fall, the new entrant would bear the economic risk of this outcome.[32] However, regulation by TSLRIC typically allows the entrant to buy the use of the unbundled element on a month-by-month basis. Thus if demand does not materialize or prices fall, the ILEC has to bear the risk for the business case of the new competitor. Accordingly, the ILEC has been required by regulation to give a free option to the new entrant, where an option is the right, but not the obligation, to purchase the use of the unbundled elements. The monthly price of the unbundled element should be significantly higher than the ten-year price of the element to reflect the risk inherent in the sunk investments, or equivalently the value of the option given to the new entrant.[33] Regulators to date, including the FCC, the ACCC in Australia and the EU have not incorporated the value of the option, which arises from the sunk cost nature of much telecommunication investment, into their price setting.

Another way to consider the problem of regulation set prices is to allow for the existence of the (all-knowing) social planner. Suppose the social planner were considering a new investment in a telecommunications network where the features of sunk and irreversible investments are important. The social planner wants to maximize the value of the social welfare integral over time subject to uncertainty. However, the investment is subject to both technological and economic uncertainty so that the cost of the investment may (randomly) decrease in the future and because of demand uncertainty the social planner does not know whether the investment will be economic. In making an optimal decision the social planner will take into account the sunk and irreversible nature of the investment since should the new service fail, the investment cannot be shifted to another use. Thus, incorrectly assuming that sunk costs do not exist – which is the perfect contestability standard – when sunk costs are an extremely important part of the economic problem, will lead to incorrect decisions and decreased economic efficiency. The economy will not reach its production possibility frontier.

Regulation Set Prices for Unbundled Elements

Under the 1996 Telecommunication Act the FCC mandated forward looking cost-based prices for competitors to use unbundled LEC facilities.[34] The FCC did not permit any mark-up over cost to allow for the risk associated with investment in sunk assets; instead, it used a TSLRIC-type approach that attempts to estimate the total service long-run incremental cost on a forward looking basis.[35] Australian and European regulators have chosen a similar approach. TSLRIC attempts to solve the perfect competition problem that price cannot equal marginal cost by allowing for the fixed costs of a given service to be recovered; that is, it allows for the recovery of the cost of investment and variable costs of providing the service over the economic lifetime of the investment. However, TSLRIC makes no allowance for the sunk and irreversible nature of telecommunications investment, so that it adopts the perfect contestability standard. The perfect contestability standard provides the incorrect economic incentives for efficient investment once technological and economic uncertainty exist. The FCC and other regulators have chosen the incorrect standard for setting regulated prices. TSLRIC will therefore lead to less innovation and decreased investment below economically efficient levels.[36]

The TSLRIC Standard and Harm to Innovation

Many new telecommunications services do not succeed. Recent failures include Picturephone services (AT&T and MCI within the past ten years) and information service gateway services offered by many ILECs. These new gateway services require substantial sunk development costs to create the large databases needed to provide information service gateways. Now when a new service is successful, under TSLRIC regulation, an ILEC competitor can purchase the service at TSLRIC. Thus for a successful new service the ILEC recovers at most its cost; for unsuccessful services it recovers nothing and loses its sunk investment. Thus the TSLRIC regulation is the analogue of a rule that would require pharmaceutical companies to sell their successful products to their generic competitors at incremental cost and would allow the pharmaceutical companies to recover their R&D and production costs on their successful new drugs, but to recover nothing on their unsuccessful attempts. This truncation of returns where a successful new telecommunications service recovers its cost (but no more), and unsuccessful new services recover nothing decreases economic incentives for innovative new services from regulated telecommunications companies. By eliminating the right tail of the distribution of returns as demonstrated in Figure 10.2, TSLRIC regulation decreases the mean of the expected return of a new project. For example, in a project with returns, y, which follow a normal distribution with mean μ and standard deviation σ, the expected value of the return when it is truncated at cost c is:

$$E(y|y<c)=\mu-\sigma M(c) \qquad (10.3)$$

where $M(c)$ is the inverse Mills ratio evaluated at c.[37] Thus the tighter the cost standard, the lower the incentives to innovate. More importantly, as

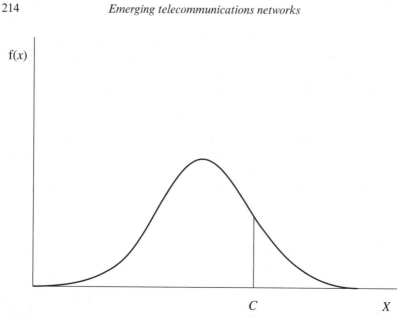

Figure 10.2　Truncated returns caused by TSLRIC

the returns to innovation become more uncertain, the expected return and the incentives to innovate decrease. Thus, in the absence of sunk and irreversible investment, a TSLRIC pricing policy decreases the economic incentives for investment in innovative services, and a TSLRIC policy may eliminate these economic incentives to invest altogether.

Regulators could allow for something similar to patent protection for new services to provide economic incentives for ILECs to innovate.[38] However, this policy option is a recipe to delay new telecommunications services for ten years or more with enormous consumer welfare losses, as occurred with voice messaging and cellular telephone.[39] Currently, it takes the US Patent Office over two years to grant a patent with longer time periods not uncommon. However, no opponent of the patent is allowed to be part of the process. In a regulatory setting where competitors would attempt to delay the introduction of new services, as happened with both voice messaging and cellular telephone, one would expect much longer delays. A better approach would be not to regulate new services. Given the large welfare gains from new services and price-cap regulation for existing services, ILECs should be permitted to offer new services with no prior approval or price regulation. The gains in consumer welfare from successful new services would lead to significant gains for consumers. Attempting to fine-tune prices of new services through cost-based regulation will lead to overall consumer losses. However, regulators find it extremely difficult not to regulate any new service of a regulated company.[40]

The Effect of Sunk and Irreversible Investments

TSLRIC assumes all capital invested now is used over the economic life of the new investment and also that prices for the capital goods or the service being offered do not decrease over time.[41] With changing demand conditions, changing prices, or changing technology, these assumptions are not necessarily true. Thus TSLRIC assumes a world of certainty where the actual world is one of uncertainty in the future. Substantial economic effects can arise from the effects that the sunk nature of investment has on the calculation of TSLRIC.

Consider the value of a project under no demand uncertainty with a risk adjusted discount rate of r and assumed known exponential economic depreciation at rate δ. This assumption on depreciation can be thought of as the price of the capital decreasing over time at this rate due to technological progress. Assume that price, net of the effect of economic depreciation of the capital goods, is expected to decrease with growth rate $-\alpha$.[42] The initial price of output is P. The value of the project is then:

$$V(P) = \int_0^\infty \lambda \exp(-\lambda t) P \frac{1 - \exp(-\delta t)}{\delta} dt = P/(\lambda + \delta) \qquad (10.4)$$

where $\lambda = r + \alpha$. Note that δ is added to the expression to account for the decreasing price of capital goods. This term, omitted from TSLRIC calculations, accounts for technological progress in equipment prices, which is one economic factor that leads to lower prices over time. Suppose that the cost of the investment is I. The rule for a competitive firm is to invest if $V(P) > I$. Equivalently from (10.4), $P > (\lambda + \delta)I$. The economic interpretation of this expression is that the price (or price minus variable cost) must exceed the cost of capital, which includes the change in price of the capital good to make the investment worthwhile.[43] Note that the net change in the output price and the price of the capital good both enter the efficient investment rule. TSLRIC calculations ignore the basic economic fact that when technological change is present, (quality adjusted) capital goods prices tend to decline over time. This economic factor needs to be taken into account or economic inefficiency will result.

A simplified example demonstrates the potential importance of changing prices of capital goods when competition exists. Suppose an investment is considered which uses computer technology in a significant manner. Because computer technology is advancing rapidly, the price of the capital good used in the investment decreases over time. Now consider a competitive firm set prices according to (10.4), but did not take account of changing prices of capital goods due to technological progress – that is, $\delta = 0$ is assumed. A company 'New Telecom' decides to enter the Internet access business. The company purchases a switch (router) which costs USD10000. It expects to serve 100 customers annually with variable costs at USD500 per annum. The firm's cost of capital is 10 per cent and it expects to use the router for five years at which time its resale (scrap) value will be zero.[44] The discounted cost of the project over five years is USD11895 which is the TSLRIC. On a per customer basis the cost is USD118.95 so that if the price were set at USD31.38 per annum the net present value (NPV) of the project is zero. Thus the price based on TSLRIC is USD31.38 per annum. Unfortunately, the company will make losses at this price and so the investment will not be made.

The reasons for this conclusion are that the price of routers, switches, fiber optic electronics and other telecommunications equipment is decreasing with technological progress – for example, Moore's law for microprocessors. Assume the price of the router declines by USD1000 annually, but all other costs remain the same. For a market entrant in year 2, the TSLRIC calculation leads to a discounted cost of USD10895 (exactly USD1000 less if no further price reductions occurred) so that the TSLRIC set price is USD28.74 per annum. Now the initial entrant, New Telecom, will be forced to decrease its price by USD2.64 and it will make a loss on every customer (taking the original cost of capital into account). Indeed, as expected, New Telecom will lose USD760 on the project. This will continue in the next year when the router price falls to USD8000. Thus TSLRIC-based prices cause the initial entrant to lose money even in a world of complete certainty because of decreasing capital costs. Instead, of charging USD31.38 for each year as TSLRIC implies, New Telecom must charge decreasing prices (USD36.65, USD33.75, USD30.85, USD27.95 and USD25.04) due to competition. Where does TSLRIC go wrong?[45] TSLRIC fails to recognize that the change in the price of the equipment needs to be included in the cost of capital. Indeed, the competitive price would not be the TSLRIC answer of USD31.38, but the correct answer is that New Telecom must charge USD36.65 in the first year and then decrease its price to USD33.75 the next year, and so on, because of the decreased price of the router. Thus the TSLRIC set price is too low by about 17 per cent for the first year because it ignores the falling price of capital goods.

The usual TSLRIC calculation does not include δ, but it instead assumes that both the prices of capital goods and the output do not change over time. This assumption is extremely inaccurate. Take a Class 5 Central Office Switch (COS) for example. Ten years ago an AT&T Class 5 switch (5-ESS) was sold to an ILEC for approximately USD200 per line (Hausman and Kohlberg, 1989; p. 204). Today, the price of Lucent 5-ESS switches and similar NTI switches are in the USD70 per line, or lower, range. A TSLRIC calculation would be based on the USD70 price. An ILEC who paid USD200 per line made the efficient investment decision when it purchased its COS. But TSLRIC, by omitting economic depreciation due to technological progress, leads to a systematically downward biased estimate of costs. Indeed the economic depreciation of COS is estimated to be near 8 per cent per annum over the past five years, while the cost of fiber optic carrier systems has decreased at approximately 7 per cent per year over the same period.[46] The omitted economic factor δ can be quite large relative to r for telecommunications switching or transmission equipment due to technological progress.

TSLRIC calculations make the further assumptions that the investment is always used at full capacity; the demand curve does not shift inwards over time; and a new or improved technology does not appear that leads to lower cost of production. Of course, these conditions are unlikely to hold for the life of the sunk investment. Thus uncertainty needs to be added to the calculation because of the sunk nature of the investment. To account for the sunk nature of the investment and its interaction with fundamental economic and technological uncertainty, assume a reward for waiting occurs because, over time, some uncertainty is resolved.[47] The uncertainty can arise from several factors: demand uncertainty; price uncertainty; technological progress (input price) uncertainty; and interest rate uncertainty.[48] Under these conditions the fundamental decision rule for investment changes to:

$$P^s > \frac{\beta_1}{\beta_1 - 1}(\delta + \lambda)I \qquad (10.5)$$

where $\beta_1 > 1$ so that $m = \beta_1/(\beta_1 - 1) > 1$. β_1 takes into account the sunk cost nature of the investment coupled with inherent economic uncertainty,[49] and m is the mark-up factor required to account for the effect of uncertain economic factors on the cost of sunk and irreversible investments. Thus the critical cut-off point for investment is $P^s > P$ from (10.2). Note that $m = 1$ for fixed but not sunk investments. Thus rearranging (10.5) gives:

$$\frac{P^s}{m} > (\delta + \lambda)I \qquad (10.6)$$

Equation (10.6) demonstrates that the value of the investment is discounted by the factor m to take account of the sunk costs, compared to the fixed (but not sunk) cost case of $m = 1$. Sunk cost investment must have higher values than fixed costs investments, other things being equal, to be economical to undertake. To see how important this consideration of sunk costs can be, next evaluate the mark-up factor m. The parameters β_1 and m depend on a number of economic factors. It can be demonstrated that as uncertainty increases; that is, the variance of the underlying stochastic process, β_1 decreases and the m factor increases (see Dixit and Pindyck, 1994, p. 153). Also, as δ increases, β_1 increases, which means that the m factor decreases. As r increases β_1 decreases, so that the m factor increases. MacDonald and Siegel (1986), and Dixit and Pindyck (1994, p. 153) calculate $m = 2$ so that, for instance, $V^S = 2I$. A TSLRIC calculation that ignores the sunk cost feature of telecommunications network investments would thus be off by a factor of two.

Using parameters for ILECs and taking account of the decrease in capital prices due to technological progress (which Dixit and Pindyck assume to be zero) and because the expected change in (real) prices of most telecommunications services is negative given the decreasing capital prices, the value of m is around 3.2 to 3.4.[50] Thus a mark-up factor must be applied to the investment cost component of TSLRIC to account for the interaction of uncertainty with sunk and irreversible costs of investment.[51] Depending on the ratio of sunk to fixed and variable costs the overall mark-up on TSLRIC will vary, but the mark-up will be substantial given the importance of sunk costs in most telecommunications investments. This mark-up would be used by a (hypothetical) social planner to determine the optimal telecommunications network investment, as the social planner faces the same inherent economic and technological uncertainty over future demand and costs.

Now when the mark-up for sunk and irreversible investment is applied, it should only be used for assets that are sunk – for example, potentially stranded. Other investments that are fixed, but not sunk, would not have the mark-up. The proportion for transport links sunk costs is 0.59 so that the mark-up factor of $m = 3.3$ is approximately 2.35 times TSLRIC. By contrast, the proportion of sunk costs for ports is about 0.10 so that the mark-up factor becomes 1.23 times TSLRIC. The mark-up over TSLRIC that takes account of sunk costs and uncertainty is the value of the free option that regulators force incumbent providers to grant to new entrants; for example, 1.35 times TSLRIC for links and 0.23 times TSLRIC for ports. Thus the proportion of sunk costs has an important effect on the correct value of regulated prices when sunk costs are taken into account. Regulators, by failing to apply a mark-up to TSLRIC, set too low a regulated price for telecommunications services from new investment. The result decreases new investment in telecommunications below economically efficient levels, contrary to the stated purpose of the 1996 Telecommunications Act and enabling legislation in other countries. Thus, through a focus on static cost efficiency considerations in setting regulated prices equal to TSLRIC, the regulators miss the negative effect on dynamic efficiency that TSLRIC-based prices cause. Since the examples of voice messaging, cellular telephone and the Internet demonstrate that the dynamic efficiency effects are quite large in telecommunications, use of TSLRIC to set regulated prices will be likely to cause substantial welfare losses to consumers similar to past FCC regulatory policy in the US.

Professor William Baumol, an inventor of contestability theory and a supporter of the TSLRIC approach to regulation, has now recognized that sunk costs must be considered in a proper regulatory approach owing to the 'profound implications for both theory and practice' (see Baumol, 1999). Because Baumol was an inventor of TSLRIC (which mutated into the TELRIC approach currently in use at the FCC) and supported the use of TSLRIC and TELRIC when the FCC decided on its current form of regulation in 1996, his recognition that sunk costs are an important economic factor that cannot be ignored is potentially quite significant.[52] Baumol now states that a cost component in the investment decision has been overlooked, so that the total costs of such decisions and hence their appropriate prices are normally underestimated. This recognition is equivalent to the granting of the free option to competitors by failing to take account of the sunk costs. Thus Baumol agrees that the options value of investment is a real cost that regulators must take account of if they are to make the correct decisions. Baumol agrees that the application of real options theory to the regulation of ILECs is potentially important, given the presence of sunk and irreversible investments. Regulators should take note of these considerations because their current TSLRIC approach assumes that sunk and irreversible investments are not present. Otherwise, regulators will be an example of Lord Keynes' (1936) observation, paraphrased in Samuelson and Nordhaus (1986), that:

> The ideas of economists and political philosophers, both when they are right and when they are wrong, are more powerful than is commonly understood, indeed the world is ruled by little else. Practical men, who believe themselves to be quite exempt from any intellectual influences are usually the slaves of some defunct [economic theory]. (p. 12)

Hopefully, regulators will realize the mistake they are making sooner rather than later.

WHAT ELEMENTS SHOULD BE UNBUNDLED?

Up to this point the choice of regulator-mandated unbundled elements, whose prices are regulated, is given exogenously. The focus has been on the correct economic method of how regulators should set prices for the elements once they are chosen. In this section the question of what elements should be unbundled is considered. If the goal is to have actual, not subsidized, competition, this choice is potentially quite important. Should regulators require essentially the entire local network to be unbundled, as the FCC has done in the US, the likely outcome is less competition. In what follows the unbundling question in the framework of the goal of consumer welfare is considered.[53] Thus the goal is not a competitor welfare goal, as regulators often seem to believe, but a consumer welfare goal. The Australian regulator, the ACCC, has explicitly established a consumer welfare goal for its approach to telecommunications regulation. The ACCC refers to the goal as the long-term interests of end users (LTIE). The FCC regulates under a public interest rule that, in my view, should be a consumer welfare rule, but the FCC has used the public interest rule to give it wide latitude in its decisions, which often have caused consumer losses of tens of billions of dollars per annum.[54]

The 1996 Telecommunications Act established basic principles for unbundling network elements. Section 251 and Section 252 provide a framework for the pricing of interconnection, resale and unbundling. Section 251(c)(3) requires any ILEC, other than certain rural carriers, to offer competitors access to the ILEC's network elements on an unbundled basis. In turn, Section 251(d)(2) requires the FCC to consider, when determining whether to mandate the unbundling of an ILEC's network elements under Section 251(c)(3), 'at a minimum, whether – access to such network elements as are proprietary in nature is necessary; and the failure to provide access to such network elements would impair the ability of the telecommunications carrier seeking access to provide the services that it seeks to offer'. Together, those subsections are known as the 'necessary' and 'impair' requirements. One cannot construe necessary and impair for purposes of Section 251(d)(2) without first identifying the larger objective of the 1996 Telecommunications Act. The statute's preamble states that its purpose is to 'promote competition and reduce regulation in order to secure lower prices and higher quality services for US telecommunications consumers and encourage the rapid deployment of new telecommunications technologies'.[55] In the legislative history, Congress reiterated that the objectives of the Telecommunications Act are 'to provide for a pro-competitive, de-regulatory national policy framework designed to accelerate rapidly private sector deployment of advanced telecommunications and information technologies and services to all Americans by opening all telecommunications markets to competition'.[56]

Consumer Welfare: Competition rather than Competitor Protection

The definitions of necessary and impair should seek to further overall competition and not merely the economic interests of individual competitors. When overall competition is increased, consumer welfare and economic efficiency will also increase. In its Local Competition First Report and Order, the FCC failed to make that distinction. Consumers benefit from competition because it leads to greater innovation and lower prices. Thus the public interest is consistent with increased competition and innovation. However, the public-interest standard, although central to interpretation of telecommunications regulation, has not always received so precise a definition in its implementation by the FCC. The primacy that economists ascribe to economic efficiency and to the maximization of consumer welfare has a related benefit. It harmonizes economic regulation and antitrust (competition) law. In 1996, Congress endorsed this view when it emphasized in the Telecommunications Act that the improvement of consumer welfare was the new legislation's overarching purpose.

A standard that looks to the effect on competition, rather than the interests of a given CLEC, complies with the US Supreme Court's command that the Commission must take into account the availability of substitutes for ILEC network elements outside the ILEC's network. When substitutes outside the ILEC's network are available, that availability occurs because some firms have made the rational economic decision that they can efficiently provide services that employ those non-ILEC elements. Two conclusions necessarily follow. The elements provided by the incumbent ILEC are not essential for competition because competition is already occurring without ILEC provision. Thus the network element, unbundled by government decree at TELRIC prices, cannot be labeled an essential facility or necessary for competition, or an element for which the decision not to mandate unbundling at a TELRIC price would impair the competitive supply of telecommunications services. Further, competition will not be adversely affected if a given CLEC cannot procure the unbundled element from the ILEC. Other firms are providing substitutes outside the ILEC's network, and so, in the absence of diminishing returns to scale, increased demand for the element outside the ILEC's network can be met at the same or lower economic cost.

The FCC's Failure to Advance Consumer Welfare

In its Local Competition First Report and Order, which it issued in 1996, the Commission determined that a requesting carrier's ability to offer service is impaired (diminished in value) if 'the quality of the service the entrant can offer absent access to the requested element, declines' or if 'the cost of providing the service rises'. That impairment standard, much like the rest of the FCC's approach to network unbundling, reflects a competitor-based standard, not a competition-based standard. The economic welfare of a single CLEC will not affect consumer welfare, because consumer welfare depends on the overall competitive supply of telecommunications services. If, under the FCC's interpretation of the necessary and impair standards, any single CLEC can claim that a given element is necessary to its business strategy, then it is likely that all elements of the network will be subjected to mandatory unbundling at TELRIC prices. Such a standard would harm consumers and diminish consumer welfare. The correct approach is for the FCC is to determine whether competition will be impaired by analyzing whether prices for telecommunications services will be higher or quality (innovation) will be lower as a result of the agency's necessary and impair policy. This approach is consistent with the ACCC LTIE standard, but is not the approach the FCC has taken.[57] Thus individual competitors' profits are not relevant to a competition standard or a public interest standard.

A Consumer Welfare Implementation of the Necessary and Impair Standard

Hausman and Sidak (1999) have proposed an approach to the necessary and impair standard of the 1996 Telecommunications Act within a consumer welfare framework. The definitions of necessary and impair rely on the competitive analysis of demand and supply substitution that provides the primary basis for other areas of regulatory economics and, more particularly, that provides the analytical basis for modern antitrust and competition law.

(i)   The essential facilities doctrine of antitrust law
The essential facilities doctrine addresses scenarios in which a company owns a resource that other firms absolutely need to provide their own services. Properly understood, the doctrine is a rule concerning the obligation (if any) of a vertically integrated firm to sell an input to competitors in the downstream market. Federal courts applied the essential facilities doctrine to telecommunications networks in *MCI Communications Corporation v.*

American Telephone & Telegraph Company.[58] In that case, the Seventh Circuit refined the essential facilities doctrines into a four-part test that requires the plaintiff to show control of the essential facility by a monopolist; a competitor's inability practically or reasonably to duplicate the essential facility; the denial of the use of the facility to a competitor; and the feasibility of providing the facility. Inherent in the concept of an essential facility is the premise that the owner of that facility possesses monopoly power. The first two elements of the doctrine incorporate that recognition in a variety of ways. Some degree of uniqueness and market control is inherent in the term 'essential'. Further, the inquiry regarding the impracticability of duplication ensures that the doctrine will apply only to facilities for which no feasible alternative exists or cannot be reasonably reproduced. Finally, the term 'facility' itself connotes an integrated physical structure or large capital asset with the degree of cost advantage or unique character that usually confers monopoly power and market control by virtue of its superiority. This approach which is applied to demonstrate the technical feasibility of access is a necessary but not sufficient condition for mandatory unbundling to advance consumer welfare.

If a given unbundled element (the facility) competes for users with other products or services that are effective substitutes for access to the facility, the discipline imposed by such competition will suffice to control the conduct of the facility owner. There will, of course, be instances in which the facility in question will be somewhat better than the alternatives, but not so much better as to preclude the continued survival of excluded parties. It may be difficult in practice to determine whether exclusion from the use of a particular facility will mean inconvenience, extinction, or some intermediate degree of harm to the excluded competitor. The point is not that the judgment as to the magnitude of the competitive disadvantage of exclusion is simpler in principle with one test instead of another. Rather, the point is that the question of essentiality and ease of duplication – measured by either the potential harm of exclusion or the potential benefit of inclusion – is no different from the issue of whether monopoly power is present in the market for the service produced with the allegedly essential facility. The focus of courts and regulators should be on whether mandatory access to the facility enhances the long-term welfare of consumers, regardless of the effect on individual competitors. Because a finding of monopoly power should be a prerequisite to any further inquiry, any market characteristic that prevents the exercise of market power should preclude the application of the essential facilities doctrine.

(ii)   Deriving the necessary and impair standards from the essential facilities doctrine

Whether the FCC should mandate unbundling of a particular network element in a particular geographic location at a particular time should depend on whether such unbundling is necessary to permit the competitive supply of telecommunications service. The correct meaning of impair for purposes of Section 251(d)(2) is whether the failure of an ILEC to unbundle a particular network element, at a TELRIC price, in a particular geographic location at a particular time would produce an equilibrium supply of telecommunications services that was, relative to the competitive equilibrium, significantly inferior for consumers. Although a particular network element may be essential to producing a bundle of services in a particular manner, the existence of competition among bundles of services limits the extent to which that element is essential to the competitive supply of telecommunications services. More specifically, the development of wireless voice, data and vertical services has served to increase the availability of substitutes for wireline access. This insight about competition at the service level is analogous to the economic concept of derived demand. In the context of Section 251(d)(2) of the Telecommunications Act, the relevant question is whether competition among bundles of services produces, for a particular network element, a sufficiently low level of derived demand such that the element is inessential to producing a competitive equilibrium.

In the language of economics, necessity and competitive impairment are given rigorous economic meaning by computing the price elasticity of derived demand for any given unbundled network element. The elasticity of derived demand for an input varies directly with Marshall's four rules of derived demand: the elasticity of demand for the product that the factor produces; the share of the factor in the cost of production; the elasticity of supply of the other factors; and the elasticity of substitution between the factor in question and the other factors. The application of Marshall's rules can illuminate whether the demand for a given network element is so inelastic – that is, the quantity demanded is not sensitive to changes in price – that it could not be considered a necessary element. The availability of close substitutes to traditional wireline service, such as wireless applications, serves to increase the elasticity of demand for wireline service and hence, by Marshall's first rule, tends to increase the elasticity of demand for all of the ILEC's network elements used to produce voice telephony. As wireless prices approach wireline prices, fixed (as opposed to mobile) customers begin to substitute wireless telephones for landline telephones. As an example, the remaining rules of derived demand are applied to loops in particular. According to Marshall's second rule, the price elasticity of derived demand for a network element should rise as the share of the element in the network costs rises. The intuition is: suppose the price of a network element, which represents a large portion of the total costs, doubles. Because the price of total network costs would rise substantially, the demand for additional network services would fall, and hence the demand for unbundled access to that particular network element would fall. An example of a network element that represents a large portion of the ILEC's total network costs is the loop. Thus Marshall's second rule implies that the price elasticity of derived demand for loops would be larger than for other network elements, *ceteris paribus*, and hence unbundled loops would be less likely to be considered necessary for competition.

According to Marshall's third rule, the price elasticity of derived demand for a loop should increase with the elasticity of supply of another network element, such as a switch. Intuitively, the more price elastic the supply of switches, the less the price of switches will fall with a given reduction in the quantity of switches employed, and hence the greater must be the reduction in the quantity of loops employed. As other network elements such as switches have become increasingly competitively supplied, Marshall's third rule of derived demand implies that the price elasticity of derived demand for loops should be rising.

Finally, according to Marshall's fourth rule, the price elasticity of derived demand for a loop should increase with an increase in the cross-price elasticity of substitution between a loop and other network elements. When network elements are used in fixed proportion, then the cross-price elasticity of substitution between a loop and another network element would be small. In that case, Marshall's fourth rule of derived demand would be the only one of the four rules that does not imply a large price elasticity of derived demand for loops. On the other hand, if technological change permits network elements to be used in variable proportions, substitution will occur across network elements, and Marshall's fourth rule of derived demand will have relevance.

(iii)   The relevant product market and critical share

The 1992 US Merger Guidelines specify that relevant markets for merger analysis may be defined for classes of customers on whom a hypothetical monopolist of the merging firms' products would be likely to impose a discriminatory price increase.[59] According to the Merger Guidelines, the task of defining the relevant product market when price discrimination is not feasible involves identifying the smallest set of products for which a hypothetical monopolist could profitably raise the price by a significant amount (typically 5 per cent) above the competitive level for a non-transitory period of time (normally assumed to be two years).[60] Thus, under the Merger Guidelines, a potential market definition is too narrow if, in the face of a 5 per cent price increase, the number of customers who would switch to products outside the market is sufficiently large to make the price increase unprofitable. Customers who decide not to purchase the product (or to purchase less of the product) at the increased price are marginal consumers. For small price increases, they switch from the products inside the putative market. Not all customers, however, are marginal customers. Indeed, in the typical case, most customers would continue to purchase the product despite the higher price because their willingness to pay for the product exceeds the raised price. These customers are infra-marginal consumers.

In the presence of high demand elasticity and high supply elasticity, a firm cannot exercise unilateral monopoly power by attempting to decrease its supply. Demand elasticity is captured by customer willingness to switch to competing suppliers as relative prices change. Thus a broad range of available substitutes would imply a high own-price elasticity of demand. Following the same logic as the market definition criteria, the Merger Guidelines provide a concrete test for evaluating the competitiveness of a market as captured in the idea of market power, which is the ability of a firm unilaterally to increase price above the competitive level for a non-transitory period.[61] Because competition takes place at the margin, only a small proportion of the ILEC's customers need to defect to defeat its attempted price increase. In a simple example, it is possible to calculate that necessary proportion. Suppose that an ILEC attempted to increase prices on end user access by 5 per cent. How much traffic would that ILEC need to lose before the increase would be unprofitable? The formula to calculate that critical share is:

$$(1 - MC/P)\, q_1 < (1.05 - MC/P)\, \dot{q}_2. \tag{10.7}$$

An important empirical fact for network elements is that fixed costs are a very large component of the overall cost, so that marginal cost is a relatively small component. Assume, for example, that the ratio of marginal cost to price, MC/P, is 0.2. Then q_2 would be $0.94q_1$, so that the critical share is 6 per cent. Thus, if the ILEC were to attempt to raise its price by 5 per cent, and if, as a result, it were to lose more than 6 per cent of its traffic, the attempted price increase would be unprofitable and thus unilaterally rescinded.[62]

(iv) The Hausman–Sidak test for the impairment standard

The existing essential facilities doctrine sets forth necessary but not sufficient conditions for defining impairment under Section 251(d)(2). The complete set of necessary and sufficient conditions includes a fifth requirement, responsive to the Telecommunications Act, to address whether the denial of access to that network element at TELRIC prices would impair competition at the end user level. The Hausman–Sidak five-part test is as follows.

The FCC should mandate unbundling of a network element if, and only if: it is technically feasible for the ILEC to provide the CLEC unbundled access to the requested network element in the relevant geographic market; the ILEC has denied the CLEC use of the network element at a regulated price; it is impractical and unreasonable for the CLEC to duplicate the requested network element through any alternative source of supply; the requested network element is controlled by an ILEC that is a monopolist in the supply of a telecommunications service to end users that employs the network element in question in the relevant geographic market; and the ILEC can exercise market power in the provision of telecommunications services to end users in the relevant geographic market by restricting access to the requested network element.

To implement the fifth element of the Hausman–Sidak test, one modifies the Merger Guideline's test for unilateral market power only slightly: whether it would impair competition for an ILEC not to sell a particular unbundled network element to a CLEC at a regulated price. Intuitively, our impairment test asks whether the ILEC can exercise market power when restricting access to a particular network element to the CLEC in a particular geographic market. If the ILEC cannot exercise market power in the output market, when declining to offer a particular network element at a TELRIC price, then all the consumer benefits associated with a competitive outcome have already been secured. Therefore, the regulator should not order the network element in question to be unbundled. In contrast to the method employed by the FCC, the Hausman–Sidak test is focused on protecting competition as opposed to competitors. Where market forces can protect consumers from the harm of monopolization, then the regulators should not impose mandatory unbundling.

Thus the answer to the question of when a network element should be unbundled is when the incumbent can exercise monopoly power in the absence of unbundling. In this situation competition is harmed and consumer welfare is decreased because consumers will pay a supra-competitive price for the final service, barring further regulatory distortions. This conclusion is very closely related to the essential insight of the economic approach to regulation. Regulation should only be used in the situation of market failure, which here would be the exercise of unilateral monopoly power. Note that the approach does not use competitor welfare as the standard – instead, consumer welfare is the appropriate standard. The approach concludes that network elements should not be unbundled nor mandatory access required when monopoly power cannot be exercised.

Competitive market forces will set the price of the elements, not regulators. Thus the economists' advice that regulated prices should be like the prices set by a competitive market leads to the conclusion that the market prices should be used, in the absence of monopoly power. While regulators typically find it difficult to let go, despite their avowals to the contrary, the market should be used to determine prices. Only when unilateral monopoly power could be exercised should unbundling be required. The presence of sunk costs is then likely to be important because it is the presence of significant sunk costs that typically is an element of barriers to entry. Thus the approach of the last section should be used. Lastly, demand conditions should be taken into account when setting the regulated prices to cover the fixed and common costs. This approach will lead to increased consumer welfare, which should be the goal of regulatory policy.

NOTES

1. However, the approach did harm consumers to a significant degree by retarding new product innovation, which is a first-order loss to economic efficiency. See Hausman (1997) for estimates of the consumer welfare loss.
2. State regulatory agencies in the US set local prices for telecommunications. California adopted price cap in 1989 and by the mid-1990s the majority of states had adopted some form of incentive regulation.
3. The Bell Operating Companies had been not allowed to provide interLATA long-distance service since the breakup of AT&T in 1984.
4. This section is based on Hausman (1997, 1999a, 1999b).
5. This section is based on Hausman and Sidak (1999).
6. Indeed, the results of such allocations depend in important ways on the units the outputs q_1 and q_2 are measured in.
7. For an example of regulators causing massive losses see Hausman (1998a), and Hausman and Shelanski (1999).
8. See Laffont and Tirole (2000) for a discussion of global price caps.
9. In practice, because of incorrect depreciation schedules and inflation, utilities often did not recover the true cost of their investments.
10. Justice Stephen Breyer of the US Supreme Court in a recent decision, AT&T Corp. v. Iowa Utilities Board, 119 S. Ct. 721 (1999), described how this outcome distorts and diminishes the actual amount of competition. Regulators are actually causing decreased competition when one of their stated goals is to increase competition.
11. For estimates of the extremely large gain to consumer welfare that can arise from new telecommunications services see Hausman (1997).
12. An early version of this type of result is in Georgescu-Roegen (1951). A textbook treatment is found in Bliss (1975, Ch. 11).
13. Indirect labor costs are embedded in the other commodity inputs used to produce a given output.
14. I do not mean to initiate or bring back hoary, and now unimportant, debates about what Marx really meant. For the reader, please do not contact me about these interpretations for I will not answer.
15. Even if labor were the only primary factor, different qualities of labor would receive different wages depending on demand conditions for the different human capital that different types of labor possess. Again the necessary conditions for the non-substitution theorems would be violated. For a further discussion see Morishima (1973).

16. Economies of scale often appear as economies of density in telecommunications, but the basic notion is the same.

17. This statement may not hold in the US in the future. The 8th Circuit Court of Appeals recently (July 2000) invalidated the FCC's approach to setting regulated prices for network elements. The Court said that in the future regulated prices must depend on actual, not hypothetical, costs. Actual costs will depend on demand. The FCC will probably attempt to evade this requirement as they have done with prior Supreme Court and Appeals Court rulings, but the FCC's future success in evasion of court directions remains uncertain.

18. For a further discussion of economies of scope with switches see Hausman and Kohlberg (1989).

19. Burmeister (1980) emphasizes the unreality of the steady-state growth assumption within labor theory of value models.

20. For a recent situation where the FCC disregarded demand conditions and caused billions of dollars in efficiency losses to the economy see Hausman (1998a). This paper demonstrates that if demand conditions had been taken into account, the efficiency losses to the economy could have been reduced to approximately zero.

21. Conference Report to the Telecommunications Act of 1996, Pub. L. No. 104-104, 110 Stat. 56.

22. FCC, 'First Report and Order, CC docket No. 96-98 and 95-185', 1 August 1996.

23. The FCC is being challenged by the ILECs in Federal Court. The US Supreme Court reversed and remanded for further consideration the FCC's regulatory approach in January 1999. See AT&T Corp. v. Iowa Utils. Bd., 119 S. Ct. 721 (1999). The key issue remanded to the FCC was what network elements should be unbundled. Justice Breyer in his separate opinion discussed the effect of the FCC approach to prices of unbundled elements and the likely negative effect on new investment and innovation in local networks, which is the subject of this chapter. In July 2000 the 8th Circuit Court of Appeals invalidated the FCC approach of basing cost estimates on a hypothetical network, rather the actual network in use. See Iowa Utils. Bd. v. FCC, No. 96-3321 (2000). The Court decision requires the FCC to modify its approach to cost estimation.

24. In considering the regulation of unbundled elements, the FCC has failed to consider whether in the absence of regulation market power could be exercised by the ILECs. Instead, the FCC has adopted a competitor welfare standard, which is inconsistent with the economic analysis of competition and the modern antitrust law. In contrast, Canadian regulators have taken competitive considerations into account in their decision on which elements should be unbundled. Hausman and Tardiff (1995) discuss competitive considerations in unbundling.

25. Economists have long agreed on this point. See Kahn (1988) for a discussion.

26. Baumol and Sidak (1994, pp. 28 and 31ff).

27. The FTC and DOJ Horizontal Merger Guidelines (1992) define a sunk cost as an 'asset that cannot be recovered through the redeployment of the asset outside the relevant market, i.e., costs uniquely incurred to supply the relevant product and geographic market' (1.32).

28. To the extent that some network elements are fixed, but not sunk, investments should not be unbundled by regulators since new entrants can enter and exit markets using these elements without undergoing sunk investments, which can create entry (and exit) barriers.

29. The electronics used in the networks need not be sunk, but much of the actual dark fiber will be a sunk investment.

30. This feature of sunk and irreversible investment has been widely recognized by economic research for over a decade (see MacDonald and Siegel, 1986). For a recent comprehensive textbook treatment see Dixit and Pindyck (1994).

31. The contestable model of competition has been highly criticized as not relating to telecommunications. Critics include Armstrong and Vickers (1995), 'In fact, of course, the industry does not remotely resemble a contestable market . . .'.

32. The contract (or regulation) could allow the new entrant to sell the use of the unbundled element to another firm if it decided to exit the industry.
33. In contracts between unregulated telecommunications companies, such as long-distance carriers, and their customers, significant discounts are given for multi-year contracts.
34. The FCC decision is currently under court appeal by the ILECs. In the FCC proceeding I provided testimony on behalf of the ILECs (see Hausman, 1996).
35. The FCC chose a variant of TSLRIC, called TELRIC for total element LRIC. However, the essential economic problem of TSLRIC also exists in TELRIC. The FCC is currently constructing a TELRIC model to be used in future regulatory proceedings.
36. TSLRIC provides the correct approach in a world with no uncertainty so long as economic depreciation is known. However, given the dynamic technological advances in telecommunications, considerable uncertainty exists, especially over the long economic lifetimes of much investment in telecommunications.
37. The inverse Mills ratio is the ratio of the density function and distribution function of the standard normal distribution evaluated at $(c - \mu)/\sigma$. The ratio increases monotonically as c decreases for given μ and σ (Greene, 1990, p. 718).
38. The FCC chief economist, Joseph Farrell (1997) considered this option.
39. See Hausman (1997) for a discussion on consumer losses from this policy.
40. The FCC, remarkably enough, has proposed to regulate new services under TSLRIC-type regulation, even when the FCC itself has found that significant competition currently exists for these services. Thus the FCC is proposing to regulate new services even when no regulation is required since no market failure exists. This unnecessary regulation is potentially extremely harmful to consumers (the public interest). See Hausman (1998a), Hausman and Shelanski (1999), and Hausman and Sidak (1999) for discussions of why regulation should consider consumer welfare to be the primary factor in public interest regulation, not the competitor welfare standard that the FCC has adopted.
41. This discussion follows Hausman (1996, 1997, 1999a, 1999b, 1999c). For papers that consider the options approach to investment in telecommunications see Alleman and Noam (1999), and Laffont and Tirole (2000).
42. This factor arises due to changes in demand and total factor productivity.
43. For simplicity assume only capital and no variable costs. Variable costs are included by interpreting P as price minus variable costs that lead to the same solution.
44. The terminal value assumption can be changed with no alteration to the conclusions.
45. TSLRIC-type formulae can be corrected by using (10.2) with δ not equal to zero to account for decreasing capital prices. However, regulators have not adopted these corrections.
46. Testimony of Prof. Jerry Hausman before the CPUC, April 1998.
47. Salinger (1998) attempts to generalize the approach of (10.4) to allow for uncertainty by appending various ad hoc assumptions on randomness to the equation. However, his approach has severe limitations. He avoids the effect of lumpy investment by assuming that investment occurs continuously, although the technological nature of much investment in telecommunications depends on its lumpiness. He also assumes that regulators update their depreciation formulae in continuous time so that the option value decreases in importance. These assumptions are similar to the contestability assumptions (instantaneous free entry and exit) that bear no relationship to the actual technology of much investment in telecommunications networks.
48. The FCC incorrectly assumed that taking account of expected price changes in capital goods and economic depreciation is sufficient to estimate the effect of changing technology and demand conditions, see the FCC 'First Report and Order', para. 686. Thus the FCC implicitly assumes that the variances of the stochastic processes that determine the uncertainty are zero, for example that no uncertainty exists. Under the FCC approach the values of all traded options should be zero (contrary to stock market fact), since the expected price change of the underlying stock does not enter the option value formula. It is the uncertainty related to the stochastic process, as well as the time to expiration, that gives value to the option as all option-pricing formulae demonstrate – for example, the Black–Scholes formula.

49. This equation is the solution to a differential equation. For a derivation see Dixit and Pindyck (1994, pp. 254–6, 279–80 and 369). The parameter β_1 depends on the expected risk adjusted discount rate of r, the expected exponential economic depreciation δ, and the net expected price $-\alpha$, and the amount of uncertainty in the underlying stochastic process. Note that this result holds under imperfect competition and other types of market structure, not just under monopoly, as some critics have claimed incorrectly. See, for example, Dixit and Pindyck (1994, Ch. 8, 'Dynamic Equilibrium in a Competitive Industry'). Imperfect competition is the expected competitive outcome in telecommunications because of the significant fixed and common costs that exist.

50. Because of the expected decrease in the price of capital goods, even if the standard deviation of the underlying stochastic process were 0.25, as high as a typical stock, the mark-up factor is 2.1. For a standard deviation 0.5, the mark-up factor is 2.4. I have also explored the effect of the finite expected economic lifetimes of the capital investments in telecommunications infrastructure. Using expected lifetimes of 10–15 years leads to only small changes in the option value formulas – for example for a project with a 12-year economic life the mark-up factor of 2.0 changes to 1.9.

51. It is the advent of competition that requires correct regulatory policy to apply the mark-up. Previously, when regulatory policy did not allow for competition, regulators could (incorrectly) set prices based on historic capital costs. Given the onset of competition arising from the 1996 Telecommunications Act and regulatory removal of barriers to competition, regulators must now account for changes in prices over time. Otherwise, ILECs will decrease their investment below economically efficient levels because their expected returns, adjusted for risk, will be too low to justify the new investment.

52. See Affidavit of Baumol, Ordover and Willig on behalf of AT&T in FCC CC Docket No. 96-98, July 1996. Also see Baumol and Gregory Sidak (1994, Ch. 6).

53. See Hausman (1998b), Hausman and Shelanski (1999), and Hausman and Sidak (1999).

54. See Hausman (1997), Hausman (1998a) and Hausman and Shelanski (1999).

55. Telecommunications Act of 1996, Pub. L. No. 104-104, pmbl., 110 Stat. 56, 56.

56. H.R. REP. NO. 104-458, p. 1 (1996).

57. In May 1997, the CRTC adopted an unbundling policy that, in contrast to the FCC's approach, ordered that Canadian ILECs 'should generally not be required to make available facilities for which there are alternative sources of supply or which (competitive local exchange carriers) can reasonably supply on their own.' Mandatory unbundling in Canada extends only to the ILEC's essential facilities.

58. 708 F.2d 1081 (7th Cir. 1983).

59. See 1992 DOJ and FTC Horizontal Merger Guidelines. The Australian 1999 Merger Guidelines take a similar approach.

60. 1992 Horizontal Merger Guidelines. For convenience, the 5 per cent, although for some purposes a 10 per cent level may be more appropriate.

61. See 1992 Horizontal Merger Guidelines. The Merger Guidelines emphasize the own-price elasticity of demand, while other analyses focus on the cross-price elasticity of demand. But the two elasticity measures are closely related.

62. For a more extensive discussion of critical share see Hausman et al. (1996).

REFERENCES

Alleman, J. and Noam, E. (eds) (1999), *The New Investment Theory of Real Options and its Implications for Telecommunications Economics*, Kluwer: Boston.

Armstrong, M. and Vickers, J. (1995), 'Regulation in telecommunications', in Bishop, M., Kay, J. and Meyer, C. (eds), *The Regulatory Challenge*, Oxford University Press: Oxford.

Baumol, W. (1999), 'Option value analysis and telephone access charges', in

Alleman, J. and Noam, E. (eds), *The New Investment Theory of Real Options and its Implications for Telecommunications Economics*, Kluwer: Boston.

Baumol, W.J. and Sidak, J.G. (1994), *Toward Competition in Local Telephony*, MIT Press: Cambridge, MA.

Beesley, M. and Littlechild, S. (1989), 'The regulation of privatized monopolies in the United Kingdom', *Rand Journal of Economics*, 20(3), 454–72.

Bliss, C. (1975), *Capital Theory and the Distribution of Income*, North-Holland: Amsterdam.

Burmeister, E. (1980), 'Critical observation on the labor theory of value and Sraffa's Standard Commodity', in Klein, L., Nerlove, M. and Tsiang, S. (eds), *Quantitative Economics and Development*, Academic Press: New York, pp. 81–103.

Dixit, A. and Pindyck, R. (1994), *Investment Under Uncertainty*, Princeton University: Press, Princeton, NJ.

Farrell, J. (1997), 'Competition, innovation and deregulation', mimeo.

Georgescu-Roegen, N. (1951), 'Some properties of a generalized Leontief Model', in Koopmans, T.C. (ed.), *Activity Analysis of Production and Allocation*, Wiley: New York.

Greene, W.H. (1990), *Econometric Analysis*, Macmillan Publishing Co: New York.

Hausman, J. (1996), 'Reply affidavit of Prof. Jerry Hausman', FCC CC Docket No. 96-98, July, mimeo.

Hausman, J. (1997), 'Valuation and the effect of regulation on new services in telecommunications', *Brookings Papers on Economic Activity: Microeconomics*.

Hausman, J. (1998a), 'Taxation by telecommunications regulation', *Tax Policy and the Economy*, 12, 29–48.

Hausman, J. (1998b), 'Telecommunications: Building the infrastructure for value creation', in Bradley, S. and Nolan, R. (eds), *Sense and Respond*, Harvard Business School Press: Boston, MA.

Hausman, J. (1999a), 'Regulation by TSLRIC: Economic effects on investment and innovation', *Multimedia Und Recht*, 8, 22–6.

Hausman, J. (1999b), 'The effect of sunk costs in telecommunication regulation', in Alleman, J. and Noam, E. (eds), *The New Investment Theory of Real Options and its Implications for Telecommunications Economics*, Kluwer: Boston.

Hausman, J. (1999c), 'Comment', in Alleman, J. and Noam, E. (eds), *The New Investment Theory of Real Options and its Implications for Telecommunications Economics*, Kluwer: Boston.

Hausman, J. and Kohlberg, W.E. (1989), 'The evolution of the central office switch industry', in Bradley, S. and Hausman, J. (eds), *Future Competition in Telecommunications*, Harvard Business School Press: Boston, MA.

Hausman, J., Leonard, G. and Velluro, C. (1996), 'Market definition under price discrimination', *Antitrust Law Journal*, 64, 367–86.

Hausman, J. and Shelanski, H. (1999), 'Economic welfare and telecommunications regulation: The e-rate policy for universal-service subsidies', *Yale Journal on Regulation*, 16(1), 19–51.

Hausman, J. and Sidak, J.G. (1999), 'A consumer-welfare approach to the mandatory unbundling of telecommunications networks', *Yale Law Journal*, 109(3), 417–505.

Hausman, J. and Tardiff, T. (1995), 'Efficient local exchange competition', *Antitrust Bulletin*, 529–56.

Kahn, A.E. (1970), *The Economics of Regulation: Economic Principles*, John Wiley & Sons, New York.

Kahn, A.E. (1988) *The Economics of Regulation*, MIT Press: Cambridge, MA.

Laffont, J.J. and Tirole, J. (2000), *Competition in Telecommunications*, MIT Press: Cambridge, MA.

MacDonald, R. and Siegel, R. (1986), 'The value of waiting to invest', *Quarterly Journal of Economics*, 101, 707–28.

Mirrlees, J. (1969), 'The dynamic nonsubstitution theorem', *Review of Economic Studies*, 36, 67–76.

Morishima, M. (1973), *Marx's Economics*, Cambridge University Press: Cambridge.

Salinger, M. (1998), 'Regulating prices to equal forward-looking costs: cost-based prices or price-based costs?', *Journal of Regulatory Economics*, 14, 149–63.

Samuelson, P.A. (1961), 'A new theorem on nonsubstitution', *Money, Growth and Methodology*, C.W.K. Gleerup: Lund, pp. 407–23.

Samuelson, P.A. (1971), 'Understanding the Marxian notion of exploitation: A summary of the so-called transformation problem between Marxian values and competitive prices', *Journal of Economic Literature*, 9, 391–431.

Samuelson, P.A. and Nordhaus, W.D. (1986), *Economics*, 12th edn, McGraw-Hill: London.

11. Universal service in the information age

Jorge Reina Schement and Scott C. Forbes

INTRODUCTION

When Theodore Vail ushered universal service into the public consciousness with the famous phrase 'one system, one policy, universal service', he set in motion a policy discourse that created the twentieth-century telephone monopoly of AT&T and connected the United States (US) into the world's foremost telecommunications system. Now on the cusp of the new century, AT&T's monopoly no longer stands, yet the vision of interconnected homes and individuals communicating with each other regardless of distance still frames this sense of global telecommunications. The telephone and its penetration level remain benchmark telecommunications measurements and still allow crude multinational comparisons of relative price, performance and efficiency levels of communications networks. However, as telecommunications choices become more personalized and begin to rely heavily on Internet-centric digital technology, the notion that universal service means only 'a telephone in every home' is antiquated and relegates countries without established telecommunications infrastructures to an inflexible development path insensitive to the potential of new technology.

All the universal services embraced and supported by US governments throughout history – from the Erie Canal and the post road, through public libraries and rural free delivery, to radio and telephone – represent attempts to fulfill the promise of democracy by enabling the political, economic and social participation of citizens. Indeed, every democracy needs an informed and involved citizenry, something possible only when its citizens have access to information about their government and the opportunity to participate in political discourse. Citizens should be able to make informed contributions and receive the benefits of the political process once they have heard a variety of opinions openly debated in the marketplace of ideas. Citizens further participate when they communicate and engage in political discourse whether local or national. For political participation to thrive in the Information Age, citizens must gain access to and effectively use the nation's information infrastructure.

When political participation defines a democracy, economic participation lends it stability. Information networks distribute economic goods and services, and add value to transactions. Networks carry information that becomes input into other products and services and transmits information that itself has value as an independent entity. As the number of participants on a network increases, so does the network's functionality and customer base; its potential increases as it adds new users. Conversely, without effective access, a person is less likely to contribute to the pool of positive effects generated from multiple interactions on the network. Thus the economic benefits of an interconnected information infrastructure accrue to the individuals on a network, to the network owners, and function as a powerful integrator for society as a whole. Communication creates society. Participation in the network forms part of the socialization process through which society engenders loyalty and avoids anomie. After all, human beings define themselves not in isolation but through contact with others. Moreover, the range of information provided by any basic telecommunications infrastructure is infinite, ranging from the mundane to the critical. Therefore, the network forms an essential structure for overcoming social fragmentation. If the US wants to encourage the sense of shared values and mutual responsibility that comes from social interaction, then maximum access to communication networks becomes a necessity.

DEFINING AND MEASURING UNIVERSAL SERVICE

At its most basic, universal service provides a system of socioeconomic support for a specific population or technology – for example, railroads, post, education, roads and canals. In traditional telecommunications policy terminology, universal service constitutes a fairly narrow set of ideas and refers to public policies whose goals are to provide households with functional telephone service. Universal service policies typically target individuals who are unable to pay for existing telecommunications services, such as isolated rural users, low-income populations and designated educational or medical institutions (Prieger, 1998). In the US, the rural and low-income groups include significant numbers of African-Americans and Hispanics. Universal service presumes individual possession of a telephone and the establishment of an appropriate account relationship with which to purchase telephone service. By contrast, universal access measures the availability of a telephone to a given individual or household without the presumption of choice. For example, offering telephone service in a country where a five-minute walk to a telephone is the established norm constitutes universal access, not universal service. Thus in universal access technology is publicly available, but the charge is levied at the point of purchase, and the user has little or no choice in determining the service.[1] Plain Old Telephone Service (POTS) is the base for US universal service policies, with all other technological options considered extensions.[2] The primary objectives plus standard and optional telephone service, as they are typically categorized within the definition of universal service in the US, are listed in Table 11.1. Standard and optional services vary by country, depending on financial constraints, political agendas and the willingness of carriers to promote penetration.[3]

Table 11.1   Universal service elements

Primary objectives	*Optional services*
Efficient billing system	Information on existing technologies
Equitable pricing	Companion services[a]
Extensive network coverage	Connection to Internet
Goal-oriented subsidization	Touchtone for tele-services
	Training programs
Standard services	*Access to public electronic databases*
Data transmission (via modem)	Advanced data transmission (e.g. ISDN)
Emergency assistance	Digital switching
Handicap access	Electronic commerce[b]
Public payphones	Video transmission
Toll and long-distance calls	
Voice transmission	

Notes:
[a] Companion services are user-oriented services native to the connecting technology that facilitate communication and information transfer; for example, voice mail, e-mail, universal messaging and Internet telephony.
[b] Electronic commerce includes online services that require the Internet and the WWW.

The Measurement of Universal Service

Many groups of telecommunications users suffer from endemically low penetration of available technologies or experience slower diffusion than the national average. The International Telecommunication Union (ITU) has cited the lack of cogent and effective universal service policies as a primary reason why there is unequal access to communications technologies:

We are profoundly concerned at the deepening mal-distribution of access, resource, and opportunities in the information and communication field. The information and technology gap and related inequities between industrialized and developing nations are widening: a new type of poverty – information poverty – looms. Most developing countries, especially the least developed countries, are not sharing in the communications revolution since they lack ... policies that promote equitable public participation in the information society as both producers and consumers of information and knowledge. (ITU, 1998a)

While individuals may lag behind majority adoption patterns due to a lack of opinionated leadership, adherence to traditional patterns of technology use or delayed market forces, it is more likely that systemic hindrance prevents the adoption of a communications system in tune with societal norms (Rogers 1995; Compaine and Weintraub, 1997). Universal service should be an active public policy that allows people to communicate – to send and receive information at a reasonable cost. From this perspective, the social value of the promise of universal service outweighs giving priority to any particular delivery mechanism, type of market, geographic location and so on. The value of universal service to many is far more than what people can afford. Often people can afford to pay very little but value the connection to their social sphere above all else. In addition, there are benefits for the providers of universal service-mandated technologies. As Xavier (1997) notes, benefits to technology suppliers and operators of implementing a universal service policy include: improving corporate image, acquiring valuable customer information, and utilizing economies of scale to reduce cost and promote a proprietary technology.

Thus measuring the effectiveness of existing telecommunications policies and the subsequent progress of a revised universal service program is critical. For example, Sawhney (1992a, 1992b) mentions that areas of low population density, regardless of gender or ethnicity, tend to have lower telephone penetration coupled with widely dispersed demand requirements: a telecommuting worker may need high-speed Internet access, while a nearby neighbor may be satisfied with a rotary telephone reminiscent of the early 1900s. Measuring this demand for service and the communications technology penetration level is key to improving the status quo. Two measures in particular are internationally recognized measurements of telephone penetration that do not focus on households: number of main telephone lines (MTL) and main telephone lines per 100 inhabitants.[4]

Only the US has more than 100 million MTLs and few markets have more than 30 million MTLs: Brazil, India, PR China, Russia and Western European members of the European Union (EU). This distribution is reassuring in that the majority of the world's population live in these locations, but equally it is disturbing as it illustrates that most nations possess fewer

than 3 million MTLS, and that even entire continents are lacking significant penetration rates – Africa has but two countries with more than 3
million MTLs. This small base of MTL-rich countries is reflected in per
capita MTL data. North American countries, with the exception of
Greenland and Mexico, have more than 60 lines per 100 persons, a 60 per
cent penetration rate. By contrast, the world per capita average is 13.3 per
cent, or 13.3 lines per 100 persons. Australia, Europe, South America and
most of Asia have penetration rates of between 40 per cent and 60 per cent.
By this measure the African continent suffers from extremely low penetration rates with an average of two MTLs per 100 persons. Perhaps more
alarming, India has a sixth of the world's population but only a 2 per cent
penetration rate.

REGIONAL COMPARISONS AND CASE STUDIES

A clear trend emerging in the telecommunications environment is the
increasingly dominant and aggressive behavior of global multi-media con-
glomerates – firms that are growing to resemble the pre-1984 AT&T. This
convergence should not come as a surprise as telecommunications, once a
disparate field of activity, has merged previously distinct industries of
telephony, commercial broadcasting, consumer electronics, and computing. Digital networks and packet switching are experiencing amounts of
streaming voice, video and data that dwarf traditional voice-only traffic.
State- or region-based pricing and operating procedures are giving way to
models emphasizing gross discounts in both the business and consumer
realm as telecommunications firms recognize the importance of distributed
networks across national borders – that is, the transmission of data across
national borders.

To survive, telecommunications companies, dominant in a single industry, must enter and succeed in the established domain of rivals. The simplest way to gain expertise in a field is to buy it and exploit economies of
scope and scale. Thus 'Bellheads' and 'Netheads' grapple for an Internet
business plan. Established customer bases and 'deep pockets' buoy up the
established players while reckless disregard for established principles,
hopeful venture capitalists, innovative designs and technology propel
entrants into the same waters as their established competitors. The telecommunications market is expanding but not as fast as market players are consolidating their positions. The completed mergers of Bell Atlantic with
Nynex coupled with the SBC and Pacific Telesis union reduced to six the
original seven Baby Bells.[5] MCI has merged with WorldCom and AT&T
was given approval by the Department of Justice (DoJ) to acquire the

largest US cable operator – TCI. In 1996, there were seven RBOCs and GTE – a supplier of telecommunications equipment and nearly equivalent in size to an RBOC. By the beginning of 1999 the number was reduced to five. With the mergers of USWest and Qwest, the union of SBC and Ameritech and the new dominance of Verizon, formerly known as Bell Atlantic, in both local and cellular telephone services, the industry continues to consolidate. Questions of public interest and reduced competition in a relevant market have been subsumed under the premise that elephants must compete with elephants. In the telecommunications world, smaller companies (mice) tend not to frighten larger competitors and they are usually bought. For example, Microsoft paid 400 million US dollars (USD) for HotMail, a growing but small free e-mail service, as it threatened Microsoft's Internet service.

A recent example of merger mania is the merging of Lucent Technologies with Ascend Communications for USD20 billion in January 1999. The merger came as little surprise to industry observers, and was intended to counteract the acquisition of Bay Networks, a supplier of Internet equipment, by Northern Telecom for USD7 billion and to prevent industry leader Cisco Systems from dominating the blossoming market for Internet routers and switches needed to keep e-mail, the WWW, and other Internet traffic flowing smoothly.[6] The move provides Lucent with an ability to produce a wider range of services, to reduce the per unit costs once reorganization is complete, and, since it greatly increased both the quality and quantity of its service offerings, to further raise the already high barriers to entry in the Internet equipment sector – a newly important telecommunications sector.

The Case of the US

In a country known for its dependence on its sophisticated telecommunications infrastructure, groups of telecommunications users continue to suffer from endemically low penetration of available technology or experience slower diffusion than the national average. The US uses telephone penetration (the percentage of households with telephone service in their homes) to measure the results of universal service policy. When all households who want telephone access have it, universal service is achieved. Households headed by ethnic minorities (excluding Asians) have telephone penetration rates substantially lower than the national average. Research also shows that ethnic minorities and young and transient persons often defer telephone service because of costs relating to use, not access. New services, such as separating local and long-distance bills, were implemented by the FCC to improve minority telephone penetration rates (Fife and

Dordick, 1991; Mueller and Schement, 1996; Schement, 1996; Schement et al., 1997). Such progress is important as the promise of universal service is empowering for underserved groups. However, results at some point must meet expectations (Schement, 1995).

The Case of the Pacific Region

According to the ITU (1998a), the Pacific Region is the fastest growing tele-communications market in the world. The annual average increase in the number of telephone lines since 1990 has surpassed all other regions.[7] This is particularly true in the case of low-income countries, which have registered growth of close to 30 per cent per annum, far exceeding that of any other developing region. This increase in MTLs is particularly surprising since mobile cellular service is the fastest growing telecommunications segment in this region. In fact, the number of cellular subscribers in the Pacific Region has nearly doubled (84 per cent increase), compared to an 11 per cent increase in fixed telephone lines. The number of Asia-Pacific cellular subscribers has increased by 40 million since 1990, and the region now accounts for a third of the global total (up from 15 per cent in 1990). For many, cellular services are the *de facto* means of connection to the public network: nearly 60 per cent of all telephone subscribers were using either fixed or mobile cellular service at May 1996 and that trend has continued to the present day.

Nevertheless, the Pacific Region's dynamism in connectivity varies greatly. Telephone penetration among the Asia-Pacific Economic Cooperation forum (APEC) members varies from 60 telephones per 100 population in the US and Canada to two per 100 in Indonesia, the Philippines and Vietnam. Yet the average MTL per 100 persons in APEC countries is nearly 28, more than double the world average (see Table 11.2). In the US, in particular, smaller household sizes and increased geographical dispersion may account for willingness of subscribers to subsidize new infrastructure (MTLs) with their monthly payments. Collectively, APEC member countries account for over 40 per cent of the world's population, and 50 per cent of MTLs and gross domestic product (GDP). Average APEC GDP per capita (approximately USD11 000) is more than double the world average (USD5000). Demographic factors are also important. Asia-Pacific immigration to the US doubled between 1980 and 1990, and persons of Asia-Pacific descent are the US's fastest growing minority. By 2020, it is expected that 20 million US residents will be of Asia-Pacific ethnic origin.[8]

Table 11.2   APEC: population, GDP and telephone density at 1997

Country	Population (million)	GDP (billion)	GDP per capita	MTLs (thousand)	MTLs 100 persons
Australia	18.6	390.9	21 348	9 350.0	50.3
Brunei	0.3	5.0	17 556	78.8	25.8
Canada	30.3	601.6	20 075	18 459.5	60.9
Chile	14.6	71.9	4 987	2 600.0	17.8
PR China	1 260.6	697.6	566	70 310.0	5.6
Hong Kong	6.5	155.1	24 578	3 646.5	56.1
Indonesia	201.4	227.4	1 146	4 982.5	2.5
Japan	126.3	4 599.7	36 546	61 525.9	48.9
Korea	46.0	484.6	10 639	20 421.9	44.4
Malaysia	21.7	99.5	4 701	4 236.3	19.6
Mexico	96.4	334.7	3 521	9 263.6	9.6
New Zealand	3.8	65.9	17 889	1 840.0	48.6
New Guinea	4.2	4.9	1 205	47.0	1.1
Peru	24.4	60.9	2 544	1 645.9	6.8
Philippines	73.5	83.5	1 162	2 078.0	2.8
Russia	147.7	440.6	2 982	26 874.6	18.2
Singapore	3.8	93.4	25 858	1 684.9	44.8
Taiwan	21.7	260.8	12 240	10 010.6	46.6
Thailand	60.6	167.5	2 820	4 815.0	7.9
US	267.9	7 636.0	28 766	170 568.2	64.3
Vietnam	76.6	23.3	310	1 587.3	2.1
Total	**2 506.7**	**16 504.8**	**11 497.1***	**426 026.5**	**27.8***

Notes:
Some data are for earlier years. GDP is measured in US dollars.
* The total is an *average* of the individual figures.

Source: Basic Telecommunications Indicators, ITU (1998b).

THEORIES AND FINDINGS

Open Competition and Choice

In principle, any entity wishing to offer a basic bundle of telecommunications goods and services should be free to enter the market. Consumers should enjoy the widest possible range of choices, in order to maximize the value of access configurations. The importance of choice derives from the convergence of previously distinct media and individualistic use of them.

In the former Bell period a form of quasi-federalism reigned. Universal service in telephony was conceived as the provision of dial tone, the mechanics of which made it difficult to imagine it in a subjective context – after all a telephone is a telephone is a telephone.[9] Yet even with POTS, communities found creative uses for the uniformly black appliance. Similarly, cable's value as a baby sitter has come to outweigh its informational content in some households (Mueller and Schement, 1996). And, as for the Internet, inventive uses propel its expansion. Indeed, inventiveness characterizes consumer behavior so much that the boundaries of these media have fallen as much from consumer imagination as they have from engineered design. Universal service should, therefore, enable individuals to invent uses that make access meaningful.

The value of choice notwithstanding, open competition as a proposal raises several interesting questions. If all providers are invited, how will they communicate their offerings to consumers? The traditional state utility commission solution was to mandate flyers to be enclosed in the telephone bill. With a convergent regulatory regime, this approach might also include the cable bill. Nonetheless, anecdotal testimony at meetings of the National Association of Regulatory and Utilities Commissioners indicate that consumers typically miss the enclosed messages. A market open to all implies marketing. Providers will feel encouraged to appeal to consumers through tactics contained in the marketer's arsenal. Markets where providers compete strongly for a share will most likely experience many pitches. While intense marketing might overwhelm consumers, it falls within the custom of US culture.

In a free-for-all market big providers are likely to squeeze out small players. Moreover, the tendency in telecommunication markets has been for the big to get bigger and so dominate. Still, if the growth of dominant corporations is a problem for policy, it should not be addressed through universal service. The purpose of universal service is to maximize access – to encumber it with the management of competition will distort its aim. Another important consideration is whether open competition guarantees service to poor, rural and/or minority households. In theory, every segment of the greater market will be served because there are profits to be made. In practice, telecommunications providers have tended to serve business and upscale household market segments. Further, even though research suggests minority households consume more advanced telephone and premier cable services than comparable white households, telecommunications marketers still tend to focus on white consumers (Mueller and Schement, 1996; Schement, 1998a). Conversely, deregulation advocates assure policymakers that, left alone, providers will eventually serve marginal market segments after highly profitable core segments are developed. They suggest emerging telecommunications markets should be compared in evolution to the diffusion of TV. That 20 million US residents remain without telephone service in the 123rd year of telephone history should counter beliefs that those without telephones are satisfied, or that service to all is inevitable. The element of open competition proposed here integrates a basic service across technology and establishes the basis for meeting a wider range of needs, but it will not in and of itself reach all segments of society. The FCC and public utility commissions (PUCs) must continue discussions aimed at meeting the needs of underserved populations.

Bundled Services

Universal service, as the enabler of basic access, should allow individuals to connect to the national network transparently across media. The bundling of telephone, broadband and Internet services enhances choice and enables consumers to tailor their preferred configuration of telecommunications services. Therefore, providers should be encouraged to offer as many bundles as they wish so as to pursue market segmentation.

Bundling services raises a fundamental question as to who will choose the bundles. Historically, the make-up of universal service reflected negotiations among the FCC, PUCs, citizen groups and the monopoly provider AT&T. For most of the century, AT&T's vision, reflected at the FCC, determined the horizon. It is possible that the FCC may decide to bundle content. To do so would lead to widespread business and consumer dissatisfaction and be out of step with the current pro-deregulation political environments found within the telecommunications sector of most industrialized countries. Alternatively, the FCC could allow providers to offer many bundles and configurations to the market. Providers would offer numerous bundles to segment the market into consumer groups that facilitate marketing and bolster brand loyalty. Given that market segmentation as a strategy is well understood by business, it can be expected that creative and aggressive bundling would occur in attempts to dominate market niches. Similarly, since US consumers are sophisticated, it can be expected that they will be discriminating in their choices. The results should be the rapid development of niches and an awareness of which population segments are overlooked.

Conversely, the lure of profits inherent in competitive bundling may lead to fraud. As 'slamming' is already a festering problem in the long-distance market, it is likely to grow even more pronounced in the bundling market.[10] When consumers face many choices, some confusion is inevitable, and conditions are ripe for unscrupulous vendors to prey on the less aware. That said, competitive bundling should be accompanied by FCC and DoJ preparedness.

Setting the Price of Bundles

Pricing possibilities include bundle providers setting their bundle prices, possibly under FCC oversight, or the FCC setting the price of a basic bundle and allowing competition on bundle contents. In the former case, even under FCC oversight, competition is fostered. However, an open-pricing regime may not provide affordable service to all US residents – for example, telephone companies have been slow to target low-income households, even though they are substantial users of advanced service. Further, entrants providing local telephone service have largely ignored the potential in low-income markets. Were the FCC to set the price of the basic bundle, the policy would offer more assurance to lower income consumers. However, this approach is likely to be opposed by vendors who would argue against being shackled to a fixed price (or prices) in such a competitive arena. Still, the relationship between affordability and access persists as a source of confusion. That is, gaining access to a network is not a problem for most low-income households; the problem is staying on the network as a result of losing control over toll charges (Mueller and Schement, 1996). The FCC could, therefore, test the feasibility of open pricing by instituting guarantees against losing basic service. Such guarantees already exist in some states for telephone access. Clearly, the challenge of affordable pricing carries complex nuances for policy-makers. Attention should be concentrated on the balance between the benefits of open competition and the necessity for guarantees that will allow lower income consumers to stay on the network.

Protection of Existing Universal Service Guarantees

As a pledge against unintentionally widening access gaps, all bundles should be required to provide existing basic telephone service at some minimum level; that is, dial tone, directory assistance, emergency assistance, local and long-distance service. There is no merit in attempting an advance by losing ground. Therefore, citizens should expect the warranty that they will not lose basic universal services in the transition to a new universal service regime. In this way, as universal service expands to embrace broadband and the Internet, it will build on its traditional base.

The Importance of Local Considerations

Agencies with an understanding of local conditions, such as state PUCs, should be encouraged to take the lead in assessing local needs in order to identify specific access needs. Recently, it has become evident that telephone penetration varies dramatically at the local level. County variations are apparent in most states, even for single ethnic and other demographic groups (Schement and Forbes, 1999; Schement, 1998b). The persistence of such findings challenges the notion that a single universal service policy offers the most effective delivery of access to all. Conditions faced by Navajo Indians on their reservation in northern Arizona vary in substantive ways from conditions faced by Latin Americans residing in Phoenix. Furthermore, it is reasonable to expect that Navajo Indians and Hispanic Americans will organize their choices according to different priorities. Recognition that such variations occur across the US in mostly unknown combinations leads to an appreciation of the importance of a universal service policy that emphasizes choice. And, for those choices to be meaningful, universal service deliberations should include state entities as part of the discourse.

Based on study findings the following lessons and questions must be addressed in providing a revised conceptualization of universal service:

(a) Individuals make choices to serve their needs by a use and gratification approach to information technology. Policy must therefore reflect the concept of varying choice and values.

(b) Seamless information environments exist and the growing use of technology and services leads to greater divergence among household use patterns.

(c) How can telecommunications policy incorporate mobility, from wireless technology and the Internet, where connectivity moves outside the household?

(d) Access to basic telephone service is a problem for the poor; constructing policies to guarantee a portion of baseline local services will be a significant challenge. Can it be assumed that local service should be included in a policy while additional services should left to individual choice?

(e) Monolithic policies will not correct local variations that exist geographically and within population segments. Therefore, local solutions should be pursued to better combat low access levels.

(f) Some poor households are substantial consumers of enhanced telephone and cable services, especially African Americans. If this is true then it is wrong for carriers to initially bypass these potential customers. Further, given that early adopters make strong contributions to the dispersion of a technology, perhaps early adopters should be subsidized through a universal service policy.

(g) Diffusion of telephone, radio, cable and VCRs is different. Some forms of technology, such as single-price devices like TV and radio, enjoy rapid initial growth while other services take longer to diffuse. Too much attention to specific devices early in their adoption pattern in a universal service policy is therefore misguided, and probably can be left to individuals and the marketplace.

The idea of monthly subscription is embedded within services, which raises the issue of whether usage-sensitive or flat rate pricing maximizes diffusion and network use. Early evidence from Europe suggests charging levels and usage-sensitive pricing have hindered the growth of Internet use and access, although recent flat rate pricing successes by UK ISPs show that limited access can quickly be reversed. Several demographic trends lend credence to the notion that the Internet can effectively connect those ignored by current technology supported by universal service.[11]

The Informed Choice Model of Universal Service

Developing a dynamic universal service policy requires that the primary principles of universal service should be identified and that any policy reached should reflect these principles. Core principles include availability, connectivity, affordability and choice. Availability focuses on mobility, thus emphasizing both the access point of a technology and the fundamental technology. Currently, LECs and ISPs offer telephone and PC-based access points to national and global networks – the Internet being the largest such network. Connectivity incorporates the notion of transparency. A communications network must be able to operate with various technologies so as to allow ease of use and viable competition among suppliers. Further, a network must be scalable to allow for future upgrades and needs. Affordability is predicated on availability and connectivity and is mostly concerned with cost, particularly access and usage cost. Usage costs are affected most by competition and regulated subsidization, and probably this will continue. Most costs depend as much on the infrastructure as on the type of telecommunications service chosen. Choice highlights the service profile of the user, and his or her willingness to pay for service options. Payment options are the core of the choice component.

The informed choice model (ICM) of universal service embraces the changing world communications landscape and recognizes that although the telephone remains the primary link for industrial households, new technology must be incorporated into the universal service concept if it is to remain a worthwhile policy goal. The ICM accommodates new telecommunications networks to circumstances experienced by the world's people. The model's emphasis on endogenous preferences – individual choice as opposed to overarching political directives – offers a standard applicable in all countries and the potential to respond to diverse national and cultural needs. This chapter has primarily relied on US data, however, lessons from the model are applicable to other countries, particularly those with large inner-city and rural communities such as India, PR China and Russia.

CONCLUSION

For the foreseeable future the emerging picture is one of telecommunications divergence with immense variations in capability. Thus service policy in the twenty-first century must respond to these differences to be considered viable. Such policy has both micro and macro aspects. At the micro level, comprehensive universal service policy can enhance access and the quality of life for populations no matter how poor, marginalized or isolated they might be. At the macro level, universal service offers a potent policy tool to advance political participation, along with the economic development of nations. Keeping both these goals in sight during public discourse over global universal service principles, society will be best positioned to judge both collective opportunities and responsibilities.

Indeed, these are complex and critical issues facing the world's population at an historical juncture. To insist on framing them solely within the constraints of the short-term needs of the corporate and governmental players is to miss an opportunity to build an equitable foundation for a global information age. Universal service should lead to free and open communications. Building a model of universal service sensitive to the varying needs of diverse populations is most likely to result in political participation, economic development and social empowerment through interaction. In the past, it was common to think of policy as statements whereby governments brought order and structure to the information environment of a particular technology. Under the old concept, universal service simply represented an intention to wire the nation. It is proposed here that welfare may be optimized if people actively choose the configuration of their access. Thus the key to an effective universal service paradigm, which can double as an effective business strategy, is to provide a menu of technology and payment choices to potential users – whether it be offering ground line telephone service, wireless PCS subscriptions, a subsidized prepaid telephone card or merely an option to pay bimonthly. It is this movement away from the static notion of universal service and toward a dynamic choice model which can initiate a reconceptualization of the global universal service discussion in the new century. Emphasizing choice over cost or type of connection is an effective method of expanding the current levels of household participation in the webs of national and international communications networks.

NOTES

1. An example of a universal access technology is a coin-operated pay phone.
2. The term POTS may give the impression of an unchanging primitive system as the base for all other enhancements; however, this is misleading. In the US, POTS has evolved from live operators placing calls individually to a completely automated system that utilizes satellite relays and computerized switches. It may be 'plain', but its technology has been ever changing.
3. 'The changing role of government in an era of telecom deregulation', Report of the Second Regulatory Colloquium held at the International Telecommunication Union, Geneva, December 1993.
4. According to the *Telecommunication Indicators Handbook*, a publication of the ITU, http://www.itu.ch, a main line is a direct exchange or shared line telephone line connecting the subscriber's terminal equipment to the public switched network and has a dedicated port in the telephone exchange equipment. This term is synonymous with the terms main station or direct exchange line that are commonly used in telecommunication documents. The number of MTLs does not measure total users as a MTL might serve several persons.
5. On 1 January 1984, AT&T was split into eight RBOCs.
6. 'Lucent deal for Ascent comes with little fanfare', *New York Times*, 14 January 1999, C1–2.
7. The Pacific region is composed of the Pacific Rim and the Asia-Pacific areas. Countries included the region are: Australia, Bangladesh, Brunei, Cambodia, China, Hong Kong, India, Indonesia, Japan, Korea, Laos, Malaysia, Mongolia, New Zealand, Philippines, Russia, Singapore, Sri Lanka, Taiwan, Thailand, and Vietnam. China and Russia sometimes do not appear on lists of Pacific Rim countries.
8. Abstracted from The Center for Pacific Rim Studies at UCLA's website, http://www.isop.ucla.edu/pacrim/.
9. For the original, see Gertrude Stein's declaration, 'Rose is a rose is a rose' in *Sacred Emily*.
10. Slamming is the illegal practice of switching a customer's long-distance carrier without full consent. It is especially prevalent in minority, immigrant and non-English speaking communities.
11. NetRatings Inc news release 8 December 1997, available at www.netratings.com/newsDec_8.htm.

REFERENCES

Compaine, B. and Weintraub, M. (1997), 'Universal access to online services: An examination of the issue', *Telecommunications Policy*, 21(1), 15–33.

Fife, M.D. and Dordick, H.S. (1991), 'Universal service in post-divestiture USA', *Telecommunications Policy*, April, 119–28.

ITU (1998a), *World Telecommunications Development Report*, Geneva, ITU.

ITU, (1998b), *Basic Telecommunications Indicators*, Geneva, ITU.

Mueller, M.L. and Schement, J.R. (1996), 'Universal service from the bottom up: A study of telephone penetration in Camden, New Jersey', *The Information Society*, 12, 273–92.

Prieger, J. (1998), 'Universal service and the Telecommunications Act of 1996', *Telecommunications Policy*, 22(1), 57–71.

Rogers, E. (1995), *Diffusion of Innovations*, 4th edn, New York, Free Press.

Sawhney, H. (1992a), 'Demand aggregation for rural telephony', *Telecommunications Policy, March*, 167–78.

Sawhney, H. (1992b), 'The public telephone network', *Telecommunications Policy*, September/October, 538–52.

Schement, J.R. (1995), 'Beyond universal service', *Telecommunications Policy*, 19(6), 477–85.

Schement, J.R. (1996), 'Toward an analysis of household information infrastructure as a dimension of the emerging information infrastructure', *Social Science Computer Review*, 14(1), 69–74.

Schement, J.R. (1998a), 'Thorough Americans: Minorities and the new media', in Garmer, A.K. (ed.) *Investing in Diversity: Advancing Opportunities for Minorities and the Media*, Washington, DC, The Aspen Institute.

Schement J.R. (1998b), 'Telephone penetration at the margins: An analysis of the period following the breakup of AT&T, 1984–1994', in Sawhney, H. and Barnett, G.A. (eds) *Progress in Communication Sciences, Volume XV: Advances in Telecommunications*, Stamford, Ablex.

Schement, J.R., Belinfante, A. and Povich, L. (1997), 'Trends in telephone penetration in the United States 1984–1994', in Noam, E.M. and Wolfson, A.J. (eds) *Globalism and Localism in Telecommunications*, Amsterdam, Elsevier Science.

Schement, J. and Forbes, S.C. (1999), 'Local dimensions of the persistent gap in telecommunications,' International Communication Association, San Francisco, May.

Xavier, P. (1997), 'Universal service and public access in the networked society', *Telecommunications Policy*, 21 (9/10), 829–43.

Index

Emerging Telecommunications Networks